P9-BIG-691

CIVILIZATION AND THE LIMPET

ALSO BY MARTIN WELLS

Brain and Behaviour in Cephalopods

Lower Animals

Octopus: Physiology and Behaviour of an Advanced Invertebrate

You and Me and the Animal World

R

Civilization and the Limpet

Martin Wells

HELIX BOOKS

PERSEUS BOOKS
Reading, Massachusetts

Library of Congress Catalog Card Number: 98-87056

ISBN 0-7382-0017-4

Perseus Books is a member of the Perseus Books Group

Jacket design by Suzanne Heiser
Text design by Karen Savary
Set in 11-point Adobe Garamond by Carlisle Communications

2 3 4 5 6 7 8 9-0201009998

Find Helix Books on the World Wide Web at
http://www.aw.com/gb/

CONTENTS

PREFACE

I am a zoologist by profession, a painter, a SCUBA diver, and a yachtsman between times. This is a book about animals. Not the sweet and furry sort, but mostly about fish and invertebrates, the worms and the squid, jellyfish and small creatures that most people prefer to ignore. It deals with things that interest me as a research worker, as a voyeur, and as a university teacher.

I have two reasons for trying to foist this rather idiosyncratic collection on you. One is simply that I believe life is more interesting if you know what is going on around you.

The second reason is that I think people ought to know something about the lives of the other inhabitants of this planet, for the good of all of us. I am a fellow of a Cambridge college and I scan the wide range of newspapers and weeklies available in our common room. Apart from a considerable space devoted to motorcars and the stock market (some poor people work for all their lives and all they ever make is money) what evidently matters to the outside world is printed or painted or presented on stage, the commentaries of people on people.

You would hardly know that other animals existed.

I find this disturbing. It implies that I am some sort of nutter, wasting my life away struggling to find out things that no normal, right-minded fellow human being wants to know. And yet we all live in a world that is changing almost daily, both technically and in the way that we think about things, as a result of biological science. It has been, for the last century and more, and the change is forever accelerating. Antibiotics and contraception, gene cloning and the green revolution, cropping the sea and conserving the forests, and the whole concept of an evolving world are issues far more important to our immediate and long-term future well-being than anything a film, or a literary critic, can say about the human condition. But we are brainwashed by the sheer volume of commentaries on commentaries. It remains a shocking fact that while most of my fellow scientists are driven to feel abashed if they have never read anything by Salinger, or can't recognize a Sickert or a Stravinsky when they see or hear one, people in the arts (and many scientists, who should know better) still show no shame in admitting to an almost total ignorance of animal biology.

This is unsettling in a democratic society. We ought at least to be informed enough to rumble the manifest nonsense sometimes fed to us by politicians, the odder greens, and the lunatic fringes of the animal rights movement. At best we might even hope to understand what we are doing to our own environment.

What I am trying to do, therefore, is to wean intelligent laypeople off an excessively cannibalistic diet of articles about *Homo sapiens* onto more mixed fare, increasing their intake of facts about other animals. I like to think that they may individually feel the better for an improvement in their diet in the short term. It could also be good for the future health of all of us. At the moment, most people miss so much and so much.*

Besides (and here ideology fades to be replaced by venality), I realise that the world is more likely to go on paying people like me if it thinks what we are doing is interesting and possibly useful. And since the taxpaying political world is not about to read any of the hundred and thirty or so highly technical articles that have constituted the lifetime end product of my research, because all that is published in the highly coded and largely inaccessible language of scientific journals, I conceive that it is part of my job to publish something of what a sample biologist thinks about in fairly normal English, which is an interesting exercise in itself. Sometimes it is only when you try to explain something without recourse to jargon that you discover how little you really understand the matter.

I should perhaps add something about the sequence and choice of subjects of the articles. I started to put this lot together during a voyage south, from Southampton in England to Banyuls, in France, in the Mediterranean, in the sailboat that we afterwards kept close to the laboratory where my wife and I worked during the summers. We would normally spend about three months in the year living aboard. *Sepiola* cost most of our capital—it was a celebration of the sons passing into gainful employment and the final paying off of the mortgage—and I hoped that I might recoup some of the running costs by writing a story

*"O fat white woman whom nobody loves. / Why do you walk through the fields in gloves. / . . . Missing so much and so much?"—Frances Cornford, British poet (1886–1960). Her husband, Francis (very confusing, this), wrote *Microcosmica academica*—in English, not Latin—which all who work in universities should read.

of the voyage out, with accounts of the sorts of animals that a yachtsman could expect to meet along the way. Hence mackerel and dolphins, basking sharks, drinking seawater, and references to sailing that crop up elsewhere. But then it just grew, and I began to feed in topics that had less and less to do with boating along the Atlantic seaboard. Finally, I ended up scrapping the sailing story altogether (serious yachtsmen would anyway regard a trip round Spain and Gibraltar, challenging enough for us amateurs, in about the same light as a picnic on the Cam), and that was a great relief because I could now insert essays on other things that have interested me. The last three articles are autobiographical, aspects of my own research (as is, as a matter of fact, the second). I have stuck throughout to marine biology, and matters arising from it, because that is the area that I know most about.

Each chapter is supposed to be independent, so you can read them in any order—this theory doesn't quite work, but it very nearly does, as I've put them in what seems to me a developing sequence. Almost any of the matters broached would be (indeed, sometimes has been) the subject of a book or books by itself. Professional biologists will recognise this, but this account is not for them; it is for you, the curious nonscientist, or the curious scientist from another discipline. I have assumed no knowledge of animals and almost no knowledge of science generally. I think these things matter. Have a nice read.

M. J. W.

SEA URCHINS

If we ever had to vote for the world's most unloved animal, the sea urchin would undoubtedly figure in the short list of most people who have met one personally.

Sea urchins have an uncanny knack of choosing to roost in exactly the place you need to put a hand or foot when clambering in or out of the water. The spines penetrate with the greatest of ease and are then almost impossible to remove, so that for the rest of the holiday you limp about, cannot sit down, and embrace your friends with a caution that can all too readily be interpreted as a lack of enthusiasm. Along the southern coasts of Europe, the sea urchin has probably spoiled more potential romances than brewer's droop.

To understand, philosophers tell us, quite untruly, is to forgive. But try, for once. What is an honest, hardworking little animal to do if the nature of its business is to browse on exposed surfaces, surrounded by pirates determined to nibble off anything painstakingly converted from algal scum into something worth eating? The poor creature is a sort of marine hedgehog, eating things nobody else cares to scavenge, and it is spiny for precisely the same reasons. It moves only slowly, it has no other defense, and its spines are a measure of the uncharitability of the world in which it lives.

Until man came along it worked well enough. Other animals observed and left well alone. But man doesn't look where he is going, and just as its terrestrial counterpart was never evolved to cope with the motorcar, so the unfortunate sea urchin was totally unprepared for the onset of an insanity that has, within the last two or three generations, compelled myriads of humans to plunge like lemmings into an environment for which they are manifestly unsuited.

It is the business of sea urchins to keep the rocks clean. Remove them, and the whole place is covered with seedling seaweeds in a matter of months. Sea urchins are largely responsible for the subtidal appearance of rocky shores, just as limpets take care of the intertidal. Each individual browses an area that can range from several square meters for the large pinkish sea urchins along the coasts of the western Atlantic seaboard to a few tens of centimeters for the small black fellows in the more crowded Mediterranean.

Unlike limpets, and hedgehogs, sea urchins can clean themselves. A limpet in a rock pool will, as often as not, be itself covered in weed, even in circumstances in which it and its neighbors have contrived to keep the whole of the rest of the area swept and tidy. There is no arrangement for mutual grooming, and the mane of weed must be a deep embarrassment for the limpet, which has a hard enough time hanging onto the rocks without all that extra drag. A hedgehog is alive with fleas that it seems to have no means of evicting. Sea urchins have no such problems. They are potentially just as attractive as settlement sites for barnacles and seaweeds and other sessile organisms, and no doubt there are plenty of marine fleas that would be delighted to take shelter if not sustenance, but they manage to fend off the colonists by a process of active antifouling that should be the envy of any yachtsman. The body surface between the spines is dotted with thousands of tiny beaks on stalks that bend over to snap at anything foolish enough to tickle the surface.

This is only part of the wonder of the surface of sea urchins. Next time you find yourself feeling uncharitable about the spines, don't simply smash every sea urchin in sight, but prize one gently off the rocks and put it in a jam jar or bowl, with enough water just to cover it. Beg, borrow, or steal a lens, because the surface of a sea urchin, close up, is a truly wondrous sight. Even without a lens you should be able to see that there is a lot more to a sea urchin than a mass of spines. Long, thin, snakelike tube feet extend between the spines and stick to the glass or to the bottom of the bowl. This is how sea urchins hang on to the rocks, and how they move about. Each tube foot is water filled, hollow and extended by pressure from within. It can bend, and wave to and fro until a contact is made, whereupon the end of the foot pulls in to make a minuscule sucker that grips the rock while the foot contracts and pulls. The pull of an individual tube foot is tiny, the pull of several hundred can hold the urchin tight to the rocks in the wave surge of a gale. It is tube feet that allow starfish—close relatives of sea urchins—to wrench open clams and oysters, working in relays until the much stronger mollusc tires.

An intriguing problem with the starfish, and indeed with the sea urchin as soon as it starts to move about, is how the activity of several hundred tube feet can possibly be coordinated, for the animals have no detectable brain. Instead, a network of nerve cells is more or less concentrated along the underside of the five radii that define the symmetry of the urchin (easy enough to see if you happen on a dead one whose spines have fallen off, but not immediately obvious in the living creature unless you happen to notice that the tube feet come in rows). The five radii converge on a ring of nervous tissue around the mouth on the underside of the animal. But that is all. No ganglia, no central control point as we are used to finding with more orthodox animals. So how is coordination achieved? How does a sea urchin decide which way to go off for a browse today, let alone find its way home afterwards, or a starfish achieve the integrated activity that will allow it to sense, seek out, dig out, and disembowel a bivalve buried in the sand?

One can get an inkling from simple experiments with the beast in a bucket. Poke the surface of the urchin gently, but repeatedly, with the tip of a pencil, or with a toothpick. The little snapping beaks and then the spines turn towards the point touched. Carry on and the effect gradually spreads, farther and farther over the surface of the sea urchin. These are local, reflex actions. Any bit of the surface will behave in the same manner. If you are unkind enough to break the beast apart—you'll have to do this anyway to get at the edible bits, of which more anon—you will find that the spines and snappers continue to work and behave in the same way in any sizable fragment.

Almost in the same way, that is. Because it does appear that there are some preferential pathways, through conduction tracts along which messages pass more rapidly than the spread across the nerve net linking the muscles that move the spines. It is difficult to show this convincingly in a sea urchin, but it works quite well in a starfish. Poke the surface as before. For a while nothing seems to happen because the snappers are not stalked in starfish and most of the spines are short and stubby. Then, quite suddenly (half a minute is sudden by starfish standards), the feet below the poke point become restless. The disturbance spreads rapidly—in some tens of seconds—along the rows of feet inward towards the mouth. And then something extraordinary begins to happen. The tube feet on all five arms act collaboratively, stepping to move the starfish steadily in one direction, away from the poking.

One might reasonably conclude that there was some sort of brain in the mouth region, but even a microscopic examination fails to reveal anything remotely resembling a brain or even an unusually concentrated section of the nerve cords running around the mouth and down the arms.

Oppure si muove. We are baffled by animals that are built like this. Trying to analyze what goes on by recording from the nerves—technically difficult because the individual nerves are so small—we bog down in a maze of reflex pathways. We are used to thinking that any sort of organization inevitably means a chain of command, a hierarchy, yet this army appears to progress by a sort of consensus, any decision is a statistical event. Perhaps we should pay more attention. A sea urchin or a starfish is a democracy (the older textbooks talk of a "republic of reflexes"), and it works. The system has a splendid invulnerability, or rather a splendid ability to survive wounding on a scale that would obliterate any other moving animal. Break up a starfish, as the oystermen used to do before they learned better, and the individual arms will crawl away; any bit of the nervous system will serve to run the show for the time being. In a little while, tiny arms will regenerate from the stump, and the starfish, now multiplied by the number of bits, is back in business. Maoist animals, as indestructible as China, starfish and sea urchins are very common creatures and have been doing very nicely since long before the dinosaurs and certainly for millions of years before the newfangled mammals began to gather illusions about which of us is actually running this planet.

The echinoderms (the sea urchins and starfish, brittle stars, and those rude objects, the sea cucumbers) owe their evident success, it would seem, not only to a unique exploitation of truly democratic principles, but also to two remarkable structural inventions. One is the matter of the skeleton. The sea urchin test, and the plates and spicules embedded in the body walls of the other echinoderms, not only confers a measure of protection, but is also a trick for building a large animal rapidly and cheaply. There is amazingly little flesh in a sea urchin. What there is is spread thinly over the spines, all over the test, inside and out, and in the form of a web of small muscles operating the jaws (break open the urchin and there is an intricate, five-sided structure, known as Aristotle's lantern—he first described it—which is the thing it chews with). Apart from this, and a spiral of gut festooned around the inside, the sea urchin is hollow. The only sizable particles of tissue, and then only in the breeding season, are the ovaries or testes, five drops of nourishment grouped around the roof of the dome. This is the only bit worth eating. Each drop, a teaspoonful from a really big urchin, tastes salty and mildly fishy. Hardly something to rely upon as a staple, but acceptable enough with brown bread and butter, washed down with an adequate supply of dry white wine.

The lack of anything worth eating is, of course, the trick. Almost everything in a sea urchin is skeleton, calcium carbonate, which is available almost for free in salt water, built into an enormous framework for what is, fleshwise, a rather small animal. Sheer size reduces the number of potential predators, and the lack of organic reward for the hard work of breaking in and chomping down all that chalk means that very few animals have evolved to crop echinoderms as a regular source of food. Echinoderms start life as minute larvae and are then as vulnerable as anybody else. But they grow at an amazing rate, spread thin on their huge frames. Barring the odd octopus and some colonial animals such as the sea fans and corals, echinoderms are the largest invertebrates that one is likely to meet.

The other secret of echinoderm success is apparently unique to the group. It is a trick that would be extraordinarily handy to ourselves if we knew how to pull it off. It relates to running costs.

The easiest way of assessing the cost of living for any animal is to measure its oxygen consumption. With few exceptions, animals are overwhelmingly aerobic. That includes echinoderms, but measure the oxygen consumption of an echinoderm and this is hard to believe. They use far less oxygen, size for size, than do other animals—even other animals of the same flesh weight. Echinoderms tick over with amazingly little expenditure of energy.

They manage this by using connective tissue rather than muscle to maintain posture. Echinoderms have a connective-tissue framework, largely composed of collagen, just as we have ourselves. Collagen threads hold the muscles together because muscle, when not actively contracting, is poor, fragile stuff. Without the connective-tissue frame, your muscles would slough off when you relaxed. If you could relax, that is. In fact, people never do. Even when asleep your muscles are working away, quietly maintaining just enough tension to keep you from going totally floppy, and as soon as you stand, or sit up, there is a great deal of activity simply to maintain posture, even if you choose not to move about. All this maintenance costs fuel, and shows up as oxygen consumption.

What the echinoderms have evolved over millions of years of unhurried existence is a means of maintaining posture without muscular activity. They have collagen that is structurally and biochemically very similar to our own. Unlike us, however, they can alter its stiffness. The collagen fibers slide over one another, leaving the body of the starfish malleable and the spines of the sea urchin free for the muscles to move about. And then, having achieved a desired posture, the whole system locks solid. The muscles can switch off, and the animal, spines and all, stays exactly as it was with no further expenditure of energy. Lock on, or relaxation, is controlled nervously, with the nerve endings releasing substances that alter the environment of the collagen fibers nearby.

The odd thing is that no other animal has tumbled to this trick. It is a great pity that we cannot manage it ourselves. At seventy my blood pressure is creeping up as the collagen in my arteries hardens with age. My connective tissue, like that of the sea urchin, changes from flexible to rigid, but it does so without the possibility of reversal and very slowly over many years. Sea urchins manage the change repeatedly and reversibly, in minutes. I resent the difference because the changes in my collagen will kill me eventually, if nothing else gets me first. So why isn't humanity—since the rich countries are full of geriatrics like myself—pouring money into research on echinoderms? It is a nice question, and I do not pretend to know the answer. Man, like every other species, has his own capabilities and his own defects. One of the latter is tunnel vision.

CIVILIZATION AND THE LIMPET

Limpets sit about, doing nothing much, most of the time. When the tide is up, or the seashore still wet enough to do so without risk of dessication, they potter off to browse the thin coating of algae growing on the rocks around them. And then, after a while, they return to the spot they left, settle down, and concentrate on digestion. Day after day. Not an exciting lifestyle, but interesting because limpets turn out to be remarkably adept at returning to exactly the place they quit several hours before. They have to be. It is the only place in the vicinity where the individual's shell exactly fits the contours of the rock. It has grown to do so precisely because the animal keeps returning to the same home. A snug fit is essential if it is to avoid drying up between tides.

So, how does the limpet do it?

And why, says the anthropocentic Philistine at my elbow, should anybody care?

Because.

Because even people who only know about people should care about limpets and their navigational problems. It will give them an insight into the rise of civilization and why people can create the things—art and literature and all that—which some people still believe to be the sole and only proper study for mankind. The essential difference between us and the limpet, and the reason this is being hacked out on a word processor by a man and not by a mollusc whose ancestors had millions of years of head start on mine, is reflected in the way in which a limpet solves its navigational problems.

Pause and consider the limpet. It must return to the place where it started, a problem common to all animals that center their lives on a home, be it a cave, a burrow, a nest, or merely a place where the creature feels secure because it knows the layout of its immediate surroundings. By what means might an animal arrange to return to a base? One can list the alternatives.

1. Look for landmarks, move from one to another, then reverse the process on the return journey.
2. The landmarks don't have to be visual. They could be auditory or chemical. The smell of home could be a satisfactory beacon.
3. Blaze a trail, creating your own landmarks, and return along that.
4. Take a bearing on a landmark, or on a skymark, sun, moon, or a star, hold that bearing and return on a reciprocal one. Skymarks are more difficult than landmarks because they move with time, and any long journey must allow for this.
5. Remember all the distances moved and turns made. Retrace, or do a little simple geometry. Dead reckoning would work fine for a limpet crawling across a rock.

Animals of one sort or another have been shown to use every one of these methods. Even celestial navigation is remarkably common, and not only among the birds and the bees; most animals seem to include some sort of internal clock. We now recognize, moreover, that birds at least, and very probably a wide variety of simpler animals, have a built-in magnetic compass, so that they can move out and back without the need of landmarks. But how does the limpet do it?

Not by celestial navigation, anyway. Limpets will home by day or night, in fog, and when covered by swirling water. They will home, moreover, if the rock that they are sitting on is rotated through 90 or 180 degrees while they are away from home—energetic students with crowbars can readily move the environment so that the sun and stars now appear in all the wrong places. The limpet plods home unperturbed.

Dead reckoning is a nonstarter, too. If getting home were to depend in any way on remembering turns and distances, simply displacing the limpet should leave it disoriented, with no means, other than chance, of finding its way home. But it doesn't. A limpet knocked off its home scar—this can be done without damage if the animal is taken by surprise when it is relaxed rather than clamped down—will find its way home at the next tide, provided—and this, as we shall see, is important—that it is placed somewhere within its normal browsing range. A limpet caught off base and displaced does not set off home on a course parallel to the true route as one would expect if it were navigating by dead reckoning or depending on a built-in compass. Homing in this animal does not seem to depend on any methodology that a yachtsman would recognize as navigational.

It could be blazing a trail, leaving a streak of mucus as it slides along on the outward journey and returning along that.

Except that it doesn't. The return path only sometimes follows the outgoing track.

Maybe it knows the local landmarks, the detailed topography of its little patch, a square meter or so around its home. Again, one can undertake simple experiments to throw the unhappy limpet off course: wait until it leaves for a browse and then whack away at the rock between limpet and home with a coal chisel, and change the topography. Or, scrub the area clean, in case the cues are chemical.

Changing the topography does, indeed, puzzle the limpet some. A groove cut in the rock between the animal and its home scar (the rock gets worn where the limpet's shell abrades the surface) stops it. The limpet casts about, moves along the side of the trench and round the end of the obstruction, carries on home. Its behavior is much the same if the home scar is covered with plaster of Paris, except that there is now no home to go to and the dispossessed animal eventually wanders off or settles down by the periphery of the obstruction. All indications are that the limpet knows exactly where home ought to be.

All this implies that the clues are chemical rather than physical. It is the only explanation that we have left. Since we know that the limpet doesn't always or even very often return along its outgoing path, we must suppose that the chemical markers that it is recognizing radiate around the home. The plaster of Paris experiment having incidentally eliminated the home beacon possibility, the most likely explanation is that the limpet is picking up trails that it laid on a succession of past excursions. There is plenty of evidence from other snails that would fit in well with such a hypothesis. Predatory snails detect and track down their snail prey, and gregarious species congregate by moving along slime trails left by their conspecifics. Airbreathing pond snails follow their own mucus tracks to return to the surface when they need to replenish the bubble of air they take with them when they dive. And so on. Most seem able to detect the direction taken by the individual that laid the trail. This is not all that surprising, because it turns out that mucus trails are highly structured, both physically and chemically. A limpet would have to distinguish its own trails from those of its neighbors, crisscrossing its own, but even that should present no great problems since each individual will produce its own protein or polysaccharide signature.

So, wait until the animal leaves home and scrub the intervening area, removing the traces of trails. This turns out to be remarkably difficult. Water won't do. Detergents and strong alkali—the sort of thing you use to unclog drains—ought to do it, but don't, not with any reliability. Scrubbing with proteases and materials that should digest and disperse other components of the mucus seem to work quite often but not always. The successes could be due to fouling the environment, as traces of the materials used to remove the trails are just about as difficult to eliminate as are the trails themselves.

So the case remains unproven. We are near enough certain that the answer lies in a highly sophisticated and very acute sense of smell—or taste, because we are dealing with contact rather than something carried by wind or water—but the clinching experiment has yet to be done.

Not, in the present context, that that matters very much. What matters are the mechanisms that the limpet *doesn't* use. There is no sign of navigation by land or skymarks. And no dead reckoning.

An ant or a bee uses skymarks. They use the sun as a compass and allow for the passing of time, so that the return course does not have to be a simple reciprocal of the angle to the sun taken on the way out. Bees can even communicate courses and distances to their sisters, so that other members of the hive can be recruited to crop a favorable food source. (Some years ago a friend, a Swiss professor, had to devise an animal behavior exhibit for an international exhibition at Lausanne. He set up a swarm of bees in a glass-fronted beehive, so that visitors were able to see the bees dancing on the comb as they returned from foraging. The angle to the vertical face of the comb indicates sun compass bearing, the frequency with which the bee waggles its abdomen gives distance. The code is quite simple, and the visitors were invited to plot off course and distance on a map provided. The bees, it turned out, were all going to the same place. They were robbing a jam factory on the far side of Lake Leman.)

So bees, like birds, can navigate. And insects, in general, seem to be quite skillful at learning to find their way about by dead reckoning. Even a cockroach will learn to run a maze, and there is no question of its laying a chemical trail, because it does just as well in a new maze of the same configuration.

For cockroaches, read crabs, or rats. But not snails, or worms, or octopuses.

So what's the difference?

Snails and the walking predatory worms that one finds in the sea (earthworms are highly specialized for their boring existence and have reduced nervous systems) would, one would think, be bright enough to cope with a simple maze—knowing its way about its environment must be useful to any animal. But they fail, dismally, compared with the arthropods, let alone the vertebrates.

This is not surprising, you may well point out. Snails and such are not renowned for intelligence; this sort of thing is probably beyond the capacity of their tiny minds.

Leaving that aside (practically *all* animals can be shown to learn, and generally quite rapidly, if you set them an appropriate task), consider the case of an undoubtedly intelligent and fast-learning relative of the snails. The octopus isn't thick, by anybody's standards. In the laboratory it can be taught visual discriminations—squares and triangles and that sort of thing—just about as rapidly as a cat or a dog, in a dozen or so trials. This is impressive when you reflect that at the start of the experiment, the octopus hasn't the foggiest idea what is expected of it. What it learns is that by coming out of its lair and grabbing one sort of shape it gets fed (usually a fragment of fish), whereas attacking the alternative results in something mildly unpleasant (a six-to-twelve-volt shock, not strong enough to persuade the animal to stay home and ignore the experiment, but enough to make it pause and consider before attacking). Experiments of this sort allow us to investigate the perceptual world of an octopus. We can discover whether it generalizes as we would (between different sizes of the same shape, for example—it does) and whether it classifies the shapes that it sees as we would (in general, it does, but there are some fascinating differences, which tell us about the construction of its brain—but this is another story and leading away from the main point).

The main point is the discriminations that the octopus *cannot* make. One would expect an animal that spends much of its life groping for food under stones or cracks in the rocks to be good at discriminating by touch. And it is—it is very good at recognizing the taste of things by touch. Those arms and suckers have a sensitivity that is well beyond that of my tongue (I know, I've tried; an octopus can distinguish differences in the concentration of tastes that we can both detect with a far greater precision than I). But it fails dramatically in any tactile discrimination that would require perception of the three-dimensional shape of an object. A sphere, for example, is indistinguishable or very nearly indistinguishable from a cube, so far as an octopus is concerned—a discrimination I find easy. And an octopus can't learn mazes.

Consider touch learning. What sources of information do we use when we want to determine the shape of an object by touch? Try it. Pick something up, with your eyes shut, and try to determine its shape. Contact with the fingertips is important, but this in itself won't allow you to determine the shape of the thing you have grasped. You have to take into account the relative position of the fingers in contact, and/or the successive positions of your fingers as you feel the object over. Knowing the position of a finger is not a matter of external stimuli, but of stimuli from within telling you about the angles of your joints. You know the relative position of parts of your own body from receptors in the joints. If you didn't have this proprioceptive sense you'd be scuppered. You could never work out the shape of an object by touch.

This is precisely the problem faced by the octopus, or, at a lower level, by the limpet or any other soft-bodied animal. Their trouble is that they lack joints. A man or a cockroach can define the position of each of its ends, fingers or feet or antennae, relative to any other bit of itself at any time. It can tell how many steps it has taken, define the angles it has turned through when making a course change, or learn the sequence of movements that it must make to carry out a task. A soft-bodied animal such as an octopus has arms and suckers that should be perfectly capable of carrying out all the manipulations that a man can manage with his fingers. But it is *too* flexible, bends in too many places, so that the task of computing where the ends are, one relative to another, although theoretically possible, is, in practice, far beyond the capacity of any reasonably sized nervous system. So there is no question of a soft, unjointed animal's learning to carry out a skilled manipulative task.

The world is thus divided into two parts. On the one hand are the soft animals: floating and crawling, swimming and burrowing, flexible animals that economize on skeletal materials, devoting nearly all their substance to the important business of reproduction. Most sorts of animal, as it turns out.

And on the other side of the great divide are the creatures with joints: ungainly animals, stomping about on legs, beings with limbs like robot limbs, extravagant in materials but capable of certain important sorts of behavior unthinkable in the more elegant soft-bodied inhabitants of our planet. A man or an insect can precisely repeat a movement, checking joint angles as it goes along. So a bee can accurately construct the hexagonal units of its honeycomb, a spider can make a web, a man weave a net, chip a flint, or build a motorcar; a rat can learn to run a complex maze, a bee can navigate and pass the information to the other members of its community; writing becomes possible. The key is measurement, measuring what one is doing with one's own body. Limpets move, but because there is no way for their soft bodies to accurately monitor the movements they have made, they can never precisely repeat a movement. Our civilization depends upon being able to do just this, on a simple, qualitative difference in the nature of the sensory information available for our brains to work on.

Thus we have two qualitatively different sorts of animal, both successful, but only one capable of manipulating its environment in a manner that has led to computers and the atom bomb. We think that this, our, sort of animal is more successful than the others, which are forever cut off from the possibility of such clever inventions. Yet we are both here in our millions, and only one of us is bashing the ozone layer.

Reflect on this next time that you meet a limpet.

ON BEING BOTH SEXES

Pity the limpet. Quite apart from the strain imposed upon it by the intellectual effort of finding its way home after any browsing expedition, a matter already discussed at some length in the last chapter, a limpet is likely to suffer severe identity crises, brought about by its sex lives. Lives, not life. For the fact of the matter is that limpets nearly all change sex as they get older. Most set up as males as soon as they are old enough to be troubled by maturity. This is no big deal. A sexually mature male limpet sits, as is the way of limpets, and does nothing, most of the time. Not for him the pursuit of nubile lady limpets. No panting scramble across the rocks, no tiny molluscan feet touching as if by magic. A limpet has nothing, or next to nothing, to fantasize about. It develops sperm and at an appropriate time of year, triggered perhaps by rising temperature and high tides, it tosses the lot off into the sea and lets the little beggars get on with it, no doubt heaving a sigh of relief that it is now all over for another year, so it can settle down to serious matters such as feeding and digestion—a vintage year for algae one can always hope—and growth.

With growth, strange things occur. As a limpet gets bigger, the male ego slides into decline, and a female personality emerges. For a while he (she) is poised indecisively on a watershed between maleness and femaleness. And then, suddenly, eggs are all the thing, spermatozoa are kids' stuff, and the now fully adult limpet settles down to be a female for the rest of its life.

Big limpets produce eggs, little limpets produce sperm. The sexes never meet, being broadcast spawners, like the overwhelming majority of invertebrate marine animals. The fertilized eggs become ciliated larvae and eventually settle, if the good Lord permits, on some suitable patch of rock, metamorphosing into microlimpets that, if the good Lord further permits (the odds against this accumulator bet are diabolical) will survive to become males and ultimately females in their turn.

"Protandrous hermaphroditism" is the zoologist's term. And very common it is, too. In terms of maximizing reproductive potential it makes a lot of sense. Sperm come cheap. They consist of little more than a warhead of DNA, the genetic code material, rocketed forward at a centimeter per hour or so by a thrashing tail, with just enough fuel to keep going for a while, targeted on any conspecific egg that happens to be in range floating in the plankton. Eggs are much more expensive to produce. An egg must contain sufficient reserves to get the little animal that is created by the act of fertilization to the point at which it can fend for itself. A good, big limpet can make plenty of eggs, a little limpet does better to devote its limited resources to sperm.

Among the many animals that pursue a protandrous strategy, another, which should be familiar to any beachcomber, is something of a celebrity. *Crepidula fornicata,* the slipper limpet (the name, I am assured by classicists, refers to the arched shape of the shell; apparently in ancient Rome prostitutes used to fornicate under the arches of the aqueducts), is interesting to biologists on several counts. It is unusual in being a filter-feeding snail. Instead of wandering about and scratching at the rocks for a living, it sits still and sucks water over its large gill, combing out the microplankton swimming in the water around it—a way of making a living that is more typical of bivalves than of the normally mobile snails. And therein lies trouble, because the animal competes with oysters. It eats the same food, sits on the same sites (hard substrates are at a premium in muddy estuaries), and given half a chance outbreeds the commercial species. Oyster farmers mutter glutinous muddy curses when the animal is mentioned. It is no comfort to European shellfishermen that they brought this blight upon themselves at the turn of the century, along with young American oysters, imported in an attempt to raise the productivity of a declining fishery.

So the animal is important economically. It also, and the two facts are not unconnected, has a curious sex life.

Crepidula's particular variant on the protandrous habit is that it only remains male if there happen to be females about. A lone larva, settling out of the plankton onto a likely patch of rock, or an oyster, which will do just as well, becomes a female. The next larva to happen along, attracted by the smell of its own kind, settles on top of the first, and becomes a male. Number three arrives and the stack builds up, forming a heap of slipper limpets. In a year or so there may be a dozen or more, founder members at the base, late arrivals joining the top of the pile. The older, larger, individuals are females, the smaller individuals males, wandering over the top of the stack and copulating with the females below. Because *Crepidula* is no broadcast spawner, it is sexually a much more complex animal than the familiar conical limpet on the rocks. A *Crepidula* male has a penis, which it uses to good effect until it has outgrown its male phase, after which this organ atrophies, the animal becomes for a while an intersex and eventually a female.

The interesting feature of the whole affair to a zoologist lies in the timing of the sex change. Any individual in isolation promptly becomes a female. So what keeps the males male? The key to the understanding of this was the observation of what happened if larvae settled downstream of a tank with a female in it. They stayed male. Something in the water from a tank with a female in it delayed their sex change.

Such "pheromones" (a name coined to describe something that has an effect like a hormone while being transmitted through the external medium rather than through the bloodstream) are now recognized as not uncommon in animals. They determine, for example, the proportions of the castes in social insects. Girls in dormitories tend to menstruate in phase. And so on. There is a lot of it about, to the delight of manufacturers of aftershaves, perfumes and other products designed to get the opposite sex upstairs.

Back, however, to the question of bisexuality. Many animals are protandrous. Others contrive to be both sexes at once. The snails in your garden carry both sets of sex organs and typically contain both eggs and sperm, a situation that necessitates very complex genitalia, since it is on the whole desirable to avoid self-fertilization. Incest is all very well if the stock is sound in the first place (the pharaohs of Egypt didn't do too badly), but it rapidly brings any genetic weakness to the fore, and it stifles variability, which is the lifeblood of adaptation should conditions change. The snails in your garden get by by producing sperm rather early in the year, before the main egg-laying season, and packaging it so that it isn't roving around the reproductive tracts when the subsequent eggs are released. Individuals seek out other similarly minded snails and swap sperm packets.

The advantage of being both sexes at once is that every meeting is a potential mating, a considerable statistical improvement on the 50:50 chance that would otherwise be the best that a randy animal could hope for.

The condition looks, on the face of it, to be a winner. And yet (let us be grateful?) it is by no means the universal practice. Simultaneous hermaphroditism (Hermes, a Greek god, male, and Aphrodite, ditto, female) is widespread only among animals that live under conditions in which meetings between the sexes are, for one reason or another, restricted. One finds it, for example, among sessile animals derived from groups that once roamed around. Barnacles are hermaphrodites; other crustaceans are not (barnacles, despite their limpetlike appearance, are related to the shrimps and lobsters, not to the molluscs). Barnacles still need to copulate to propagate their kind, but even the most spectacular penis can only reach as far as a few of the neighbors. For much the same reason, parasitic animals are typically hermaphrodites. The chances of one tapeworm's becoming established in the gut of a suitable host are small; the chances of two tapeworms doing so are smaller still. Self-fertilization is here a fallback position if nothing better offers. Lineages of tapeworms with hermaphrodite tendencies are plainly at an advantage, since any two colonists can interbreed.

The same applies even to free-living animals, if they are obliged to exploit a patchy food source. Many of the very beautiful nudibranch sea slugs are extreme specialists, feeding only on particular types of sea anemones. Their specialized tastes are related to their ability to ingest this unpromising food without setting off the stinging cells that normally protect the prey. Nudibranchs settle in the vicinity of their food species as larvae and have only very limited mobility thereafter. The tapeworm situation applies. For a reasonable chance of reproduction, every meeting should be a mating; they are all hermaphrodites.

Terrestrial animals descended from marine ancestors are often ill equipped to locate the opposite sex at a distance because the chemical senses inherited from their forebears do not work as well in air as underwater. Marine snails are rarely hermaphrodite. They find each other by smell, working upcurrent to locate members of the opposite sex, which advertise their presence by dropping scent into the water. Substances that carry well enough in water are rarely suitable for airborne communication, and, anyway, air wafts about rather too quickly and rather too variably for a slow-moving animal like a slug to stand much of a chance of meeting a mate by crawling upwind. Earthworms and leeches, hermaphrodites all, have a similar history and similar problems. Their marine relatives nearly all enjoy separate sexes.

Freshwater animals, one might think, have all the opportunities of their marine relatives and could therefore be expected to conform to the normal, bisexual, pattern. The fact that many of them do not seems odd, until one realizes that many of them, indeed most in temperate climates, have come to live in freshwater by the overland route rather than by ascending the rivers. Most of the snails that you find in freshwater are pulmonates, air-breathing relatives of the slugs and snails eating the vegetables in your garden. The worms, including leeches, are all descended from earthworms, not the walking and swimming ragworms of the sea. For the fact is that living in freshwater is physiologically difficult for any animal not enclosed in a waterproof covering.

Rainwater is distilled water, and animals are full of salts and proteins that tend to draw the stuff in through their permeable skins. The cost of living in freshwater is continuous bailing, a metabolically expensive fight against inflation; without bailing, a freshwater animal would simply swell up and burst. Evidently it has proved easier in the past to solve the problem of dessication first and then, having evolved relatively impermeable skins to keep the water in, plunge back into an environment where the problem is how to keep it out. Hence, *inter alia,* the success of the insects as freshwater animals. Insects carry a thin layer of wax on the outsides of their cuticles, so that an insect, or an insect larva, in freshwater doesn't really get wet at all. The only water coming aboard is what it chooses to drink.

The hermaphrodite option might, in some circumstances, be a better proposition anyway. Freshwater pools are often isolated, sometimes temporary. An early colonist might well be a loner, or one of a very small number of colonists that somehow made it from one puddle to the next. The problem of meeting a member of the opposite sex recurs. As indeed it does on islands, which may be thought of as ponds of land separated by large tracts of ocean. Here, too, a hermaphrodite species has a greater chance of establishing itself; land snails are great colonizers.

But all these—islands, ponds, other people's insides, and other patchy and potentially isolated environments—are special cases. In the normal manner of things, the overwhelming majority of animals have no great difficulty meeting other members of their own species, and the rule is that animals come either male or female, but not both, either successively or simultaneously.

Why? Hermaphroditism would seem to have a lot going for it. Every meeting a mating. Every colonist, or pair of colonists, a potential population. Why aren't all animals built this way? Come to that, why have two sexes at all, when one would seem to be entirely sufficient? Down with males anyway, the parasitic sex.

This is not the place to discuss the crazier aspects of feminism, except to point out that a number of animals have quite successfully abandoned sex in favor of direct development of unfertilized eggs. These, some would say enlightened, beings are scattered throughout the animal kingdom, from rotifers to reptiles (males are quite unknown in several species of lizard), but they are rare. Two sexes is the norm, even though in the majority of cases the male contributes nothing to the upbringing of the young.

The success of the two-sex system arises because families with two sexes are more variable than those that do not enjoy this state of affairs. A single unfertilized parent can only produce replicates of herself. If two individuals contribute to the genetic makeup of the next generation, combinations of the good and bad characteristics of both parents will turn up in the offspring. Rough on the kids who drew the worst of both parents in the lottery, great for the winners inheriting the best. Every animal that ever was (with the exception of the phoenix, which is, as you will note, now extinct) has produced more offspring than necessary for simple replacement. The losers lose, the winners are the fathers and mothers of the next generation. In any but a very stable world, variability will increase the odds that somebody will survive when conditions change. In a very stable world, of course, the all-female situation pays off, provided the stock is nicely adapted in the first place. It is a very good way of building up a large population fast, and it has all the advantages of hermaphroditism when colonizing a patchy environment. The aphids on your rosebushes are witness to that. But then aphids, as you might have guessed, have the best of both worlds. After a few dozen generations as females, just when the shattered roses are giving up in the autumn, aphids revert to a bisexual mode, sprout wings and leave to hibernate elsewhere. A considerable number of invertebrate animals pull this trick, throwing in a few asexual generations between periods of sexuality. Very few have abandoned the bisexual mode altogether, the very rarity of such lines implying that sooner or later unisexual opportunism is caught out by changes to the environment.

So the rule is two sexes, and cross-fertilization. But why two individuals as well as two sexes? Hermaphroditism is widespread, and it is plainly possible with a little mechanical ingenuity to avoid self-fertilization. Why aren't all animals hermaphrodite?

One reason is economics. Being both sexes at once means having two sets of sexual apparatus, oviducts and sperm ducts, brood pouches and penises, and all the rest of it. All that paraphernalia absorbs resources that could otherwise be devoted to the production of more eggs and sperm, more potential individuals in the next generation. An animal's body is, in the final analysis, no more than a mechanism for the production of further individuals of like type. This is the only capacity that matters; other things being equal, the prolific inherit the earth. It is only worth investing resources in the rest of the apparatus of reproduction if it increases the number of offspring that survive to reproduce in their turn.

A second reason is that sexual specialists will almost inevitably win out in a world of hermaphrodites. Consider what must happen in a population of hermaphrodite animals if a single sexual specialist, say, a male, is produced by some freak of genetics. The freak, uncluttered by female machinery, makes more sperm than his heterosexual neighbors. If they are broadcast spawners, chance will increase the odds that he fathers more than his fair share of the next generation. If copulation is in the nature of the beast, the specialist male wins out again, being able to devote his full energies to the matter. His offspring will inherit this apparently perverted trait. So the perversion spreads, and eventually hermaphrodites bearing female genitalia become rarer than the specialist males. The best chance of passing on your own particular genes now lies in eggs. Selection will favor specialist females, the male part of hermaphrodite individuals now standing rather little chance, heavily outnumbered by the male specialists. The situation oscillates, but will eventually settle down, other things being equal (again!) to a 50:50 ratio between male and female specialists.

Thank goodness.

My mate and I have two children, both male specialists, enough to replace the genes, something I personally care about. So far (1998), our children have spawned three grandchildren. Nobody is coming or intending to come even close to achieving their biological potential. But morals have now entered into reproduction, for the first time ever in the animal world, and responsible members of our species (we have to exclude the pope, for example) are obliged to conclude that it is no longer proper to compete in the reproductive race. We shall ruin the planet if we do.

But that is another story, not a matter of concern to limpets.

HOW OLD IS A FISH?

When we first sailed our boat out to the Mediterranean, we stopped at Gibraltar to pick up fuel, duty-free liquor, and mail at the Royal Gibraltar Yacht Club. John Bull's other island, more British than the British; they play cricket on a clay pitch and stock Marmite and Lipton's tea in the shops. Smart men in Bermuda shorts met us on the quayside and handed us a copy of Gibraltar pratique note 30 (3): "All ships alongside shall affix efficient rat guards on every line or wire connected to or reaching the shore." I had never seen a sailboat with rat guards, but it was nice of them to treat us as a serious item of shipping. We don't, of course, have any rat guards, or even a mousetrap. So far this has not worried us, though I wonder what we should actually do if invaded. I guess that a modern fiberglass yacht is not really a desirable residence for a rat. The bulkheads are impenetrable, the fiberglass dangerous if ingested, and most of the food is in the cold box or in tins. Hard life for a rat; one hopes it would take a quick look round and walk back down one of the lines or wires connected to or reaching the shore.*

A mouse in a yacht, however, might just manage on the crumbs, and then we would have a dilemma. I have a lot of sympathy for mice. Humans are curiously soppy about wee, cowering, timorous beasties, and I don't like killing things unnecessarily. I should probably try to catch it and put it ashore. Whether that would in fact be a kindness is debatable; that is the dilemma. Mice are hideously xenophobic, and the mouse's future, particularly if a male mouse (as is likely, since it would probably be a refugee from some territorial dispute in the first place), would be bleak.

The human in me says, "Give it a fighting chance." But it probably came aboard in the first place because it was a loser. Perhaps I should fatten it up before release. A really kind soul would lace its diet with steroids: "At least it will be quick, all over one way or another in twenty-four hours." The human forgets we are dealing with a mouse. The one thing that is quite certain is that it won't be quick. Not for a mouse. A mouse has a life expectancy of something under two years, while I am still more or less functional at threescore and ten. Twenty-four hours is a long haul in the life of a mouse, the equivalent of something between a week and a month in mine. During that time it will be battered and bitten to death by the local mice defending their territories, hunger

*Don't you believe it. Two years after I wrote this we picked up a rat in Corsica. I think it came off a yacht moored alongside, fresh from the Caribbean. We could hear it gnawing in the night. In the end my wife caught it by putting a wedge under the lid of the waste bin, with a string to her bunk. Heard the rat in the bin, pulled the string. In the morning I swam the bin ashore and released the rat. Rat now lives, I hope—it was rather a nice young rat—in Sardinia. This is how you spread plague.

and thirst interminably, cower and run. And run. During the twenty-four hours following its liberation, supposing that it survives that long, its little heart will have beat something like one million times, more than ten times as many beats as mine. It will have taken ten times as many breaths, and to fuel all this it must somehow find ten times as much food for every gram of its body. Mice live quicker than we do.

And mice are no exception. Small mammals always run through their lives at a pace—in terms of footfalls if not furlongs—that would leave their larger relatives literally breathless. One can, in fact, be quite precise about it. The formula is: specific fuel consumption (food needed per gram per hour) falls with increasing body weight as $\text{Weight}^{-0.25}$ (double the weight, metabolic rate rises by 84 percent, not 100 percent; at ten times the weight, the cost per gram is down to 56 percent). The $\text{Weight}^{-0.25}$ exponent applies to practically any activity you care to think about: heartbeats or breaths, time to digest a meal or reach sexual maturity, gestation and lactation periods, life expectancy. It means that all of us mammals get through just about the same number of heartbeats or square meals in the course of a lifetime. It is one of the universal laws regulating the life of mammals, and, amazingly, we have no clear idea why it should be.

Whatever the cause, it has some rather curious consequences, because all of us, of whatever size and rate of tick-over, inhabit the same planet, so that, like it or not, a day-night cycle lasts twenty-four hours, a tidal cycle runs over twenty-eight days, and seasons happen yearly. A twelve-hour night is inconveniently long for small mammals and for birds, which have to pack away enough food inside them to see them through what is, for them, a horribly long fast. Small birds in Europe or North America only just about make it in summer, when the days are long and the nights short. In the winter, most clear out and fly south. The very smallest, hummingbirds, couldn't even make it in summer in northern latitudes if they didn't go torpid and drop their body temperature at night, slowing down physiological time so that their reserves last until morning. Bats have the same trouble with days, and the smaller ones adopt the same solution. In winter, bats in temperate climates huddle together in some secluded roost and maintain a body temperature that is just sufficient to stop them from freezing. With a bit of luck their fat reserves will last them until the insects come out again in the spring.

Shrews and many small mice just keep going, wake up and ferret around for food every few minutes. Most stash away supplies so that there are snacks available whenever they wake up. At the other end of the scale one might imagine that an elephant or a hippopotamus would appreciate a significant slowing of astronomical time.

(Humans are unusual in this as in so many other ways, a difficult animal to fit into the general picture. One of our many outstanding features is that we live too long, by the standards of other mammals. Placed on the shrew-to-elephant weight scale, we fit in nicely in terms of metabolic rate, but ought to die off in our thirties; we've had our ration of heartbeats and square meals by then. The fact that we somehow contrive to live for twice as long is, arguably, a matter of brains, and looking after ourselves properly. But even animals in zoos, which typically live for a lot longer than their relatives in the wild, don't manage to beat the system by a factor of two or three as we do, so that the notion that we get by by coddling ourselves is unlikely to be the whole explanation.)

The effect of all this on marine animals depends upon whom we are discussing. Marine mammals, understandably, are all big. Anything under about the size of a baby dolphin would cool off too quickly (large surface-to-volume ratio, and the high thermal conductivity of water), and they are all in the diving business, where it pays to have a relatively low metabolic rate. The birds at sea, too, tend to be on the big side, not so much because of cooling—though that could be a serious problem on a cold and windy night—as because they need the range; fat reserves depend on the size of the body, while the cost per gram of flying from place to place, for a whole variety of reasons, drops with increasing size. With no shelter, besides, it is important to be able to fly fast, and a small bird is blown backwards in a gale.

For cold-blooded animals and those of us who study them, physiological time becomes a more complex issue. Animals are driven by biochemical machinery regulated by enzyme catalysts, each of which has an optimal temperature of operation. The enzyme complement of a cold-water species has been selected for its capacity to remain active at temperatures that would freeze the enzymes of a visitor from the tropics into almost total inactivity, and vice versa. As a result, there is a limited temperature range within which each species, or each population of a widespread species, can live. Within that range, temperature change can produce very large changes in metabolic rate since, broadly speaking, enzymatic reactions, like other chemical processes, double with each ten-degree-Celsius rise in temperature. So the seasonal difference in the rate of living of a cold-blooded animal can be even more spectacular than the change in the rate of living of a hibernating bat or dormouse. The mackerel that dashes about frenetically throughout most of the summer can drop to cold, deep water and do practically nothing all winter without running itself out of reserves. Its low metabolic rate has also, of course, cut down dramatically on its rate of aging. In physiological terms, it is almost meaningless to describe the age of a fish in months or years.

The change of gear that occurs when a fish warms up is further enhanced when it begins to feed. Digestion costs; starvation is cheap if you are not obliged to keep on burning fuel in order to maintain body heat. Warm-blooded animals expend such a high proportion of their fuel on central heating that the difference between feeding and starvation is really comparatively trivial. A heavy meal slows us down a bit and induces a pleasant drowsiness, but scarcely alters our metabolic rate; it has practically no effect whatever on the rate of energy expenditure of a mouse or a small bird. Not so for the cold-blooded. Here the basal rate is so low compared with that of a homeotherm (a warm-blooded temperature-controlled bird or mammal) that a square meal can well double the metabolic rate. Regular feeding thus imposes a further gear change on top of the accelerating effect of a rise in temperature. Cold-blooded animals light up and shut down to an extent inconceivable in a homeotherm. An ectotherm (anything that does not control its body temperature by burning fuel as do birds and mammals) can afford to be an opportunist, feeding heartily in times of plenty, conserving energy by doing nothing when the economics of seeking food become marginal or negative. An ectotherm may be hungry, but hunger will never drive it to the levels of desperation of a starving homeotherm, which must eat or cool down and die. It is no cruelty to leave your goldfish without food while you go on holiday. At worst, it will be a little leaner when you get back if the weather is hot while you are away. Indeed, it is probably better to let it starve for a bit. As its metabolism slows, it will excrete less waste and the water is less likely to go foul.

Biologists who study the economy of the oceans, and people who want to farm fish profitably, have to come to terms with these great changes in the rate of living of their subjects. The changes in an ecosystem may be seasonal, or quite local, a warm current here, or a cold upwelling there. The animals may migrate vertically, rising through a temperature gradient to feed at the surface during the day or night. And because each species has a temperature optimum, they may not all be behaving in the same way at the same time. Fisheries ecology is a beast of a subject at the best of times. When you impose upon it the inept interference that politicians can inject into even a simple scientific subject, it is a wonder that fisheries scientists remain sane at all. The subject is almost as impossible as economics—one reason why I stick to being a physiologist.

HOT FISH

Our lot, the mammals, are warm-blooded. So are the birds. We both maintain a high constant body temperature. Other animals are cold-blooded, or so the teaching goes. The teaching is, of course, oversimplified—teaching nearly always is.

For one thing, not all birds and mammals *do* maintain a constant high temperature. Bats lower theirs during the day, and hummingbirds do so at night. Without this, the central heating bills of these small animals—too much body surface radiating heat and too little body volume to store fuel—would be too great, the stores gathered during the day or night scarcely sufficient to see them through to the next feeding period. And they must do better than that if they are to stockpile the fat necessary to migrate or sit out the winter. Larger mammals work on a longer time scale, but many of these, too, curl up and cool down, stretching their resources to last longer as they hibernate.

More interesting and perhaps less universally recognized is that many of the animals normally thought of as cold-blooded are, in practice, quite as hot inside as we are, though, again, perhaps not all of the time. Flying insects expend energy at an awesome rate, being aerodynamically disastrous and possessed of remarkably inefficient muscles (another story, but leave that for now). They work very hard to stay airborne and get very hot inside. A flying bumblebee is as warm as a running human.

It has to be. The power output from a muscle depends upon its cross section and how often it can contract. The force that a muscle can exert per square centimeter of cross section is near enough constant, clear across the board (it is about four kilograms). But like all chemical processes, the changes taking place during the contraction of a muscle are temperature limited, the hotter the faster. Power output depends on the work (force times distance) that a muscle can do and how often it can do it. Many insects simply cannot contract their wing muscles rapidly enough to produce the power to sustain flight until they have warmed themselves up to something approaching our own body temperature. One can see the consequences of this in moths, for example. On a cold day, or before taking off at night, many of them sit and shudder, flexing the flight muscles without actually beating the wings until they have worked up enough heat to take off.

Something of the same difficulty faces reptiles and other cold-blooded animals. Snakes and lizards in temperate climates bask in the sun, warming themselves to levels well into the warm-blooded range before they set out to find food or risk exposure where they may have to make a run for it. The nervous system, like everything else, is temperature dependent; a hot lizard is alert and fast thinking and can move like greased lightning, the equal of anybody around until it cools down at night and again becomes sluggish.

That, you might think, was an irredeemable fault in the design. The mammal should always get the reptile in the end, just wait until it cools off. But there are still lots of reptiles about, and not only in hot climates, though they naturally do better there. The advantage of solar heating is that it comes free. A reptile can survive on a fraction of the food needed by a bird or a mammal, in which something like four-fifths of the daily intake is burned off keeping the body warm. It allows the cold-blooded to survive long periods with nothing to eat, and it allows populations to achieve much higher densities than are possible for the populations of their endothermic relatives. A snake or a lizard or a crocodile can live year long on the fat accumulated during a brief seasonal gorging, whereas similar-sized mammals would starve to death searching for food that has decamped or shriveled up. The race is not always to the perpetually swift; it is sometimes useful to be able to change gear and lie low for a while.

Fish, though, were the subject of this essay. "A cold fish," we say when we want to be mildly abusive to a fellow mammal. But some fish are not. Some, surprisingly, are almost as warm inside as a flying insect or a mammal. These are the fast fish, the elite athletes of the sea, tuna and their close relatives. And a few of the sharks. Predators all, swimming faster because their muscles are warmer than those of their prey.

Freshly caught tuna are warm to the touch. The scientific world seems to have first been alerted to this fact by one Davy, assistant inspector of army hospitals, in a paper delivered to the Royal Society in 1835. He took the temperature of a fresh-caught bonito and observed that it was 99 degrees Fahrenheit, while the sea was only at 80.5. He pointed out that bonito and other tunny had very big hearts and dark red muscle. He said that his opinion that such fish were warm-blooded "was corroborated by the testimony of several intelligent fishermen."

This was a remarkable observation. Fish have gills. The entire blood supply passes through the gills of a fish once on every circuit. Gill membranes are very thin, as they have to be if oxygen is to pass from the water into the blood during the second or so that the blood spends in the gill capillaries. If oxygen can leak in, heat can leak out. Thermal diffusion is about ten times as fast as molecular diffusion. A fish cannot be warm-blooded; the gills are far too efficient as a heat sink.

So how can a tuna sustain a temperature in its muscles well in excess of the temperature of the sea in which it is swimming? Most fish are not so gifted. What have tuna got that the rest have not? The answer is structure: a beautifully engineered heat-exchange system that traps heat where it is needed, in the muscles. It is yet another variant on the lunatic plumber—the countercurrent exchange system to be discussed in relation to fish swim bladders in "Buoyancy." Only this time heat, not oxygen, is trapped at the end of a system of parallel blood pipes, the arteries running into the muscles intimately mixed with the veins coming out; heat, remember, transfers more easily than oxygen.

The tuna does not, as a matter of fact, keep all its muscles hot. Hot muscle delivers more power, but consumes more oxygen in doing so than the cold stuff. Tuna have unusually extensive gills, but even these cannot possibly extract enough oxygen to fuel all of the muscles all of the time. The tuna only keeps hot the 20 percent or so of its muscles that it uses for continuous fast swimming. The rest is held in reserve, a powerhouse available for the occasional quick sprint that may make a life-or-death difference to the tuna or to its prey. This is the normal situation in fish; the brown strip along the flank of a herring or a mackerel is the only bit that the fish uses for routine swimming. The reason fish flesh is typically white rather than red is that the white bits contain few blood vessels; there isn't enough oxygen to go round, and the fish makes no attempt to keep it supplied on a minute-to-minute basis. When the fish does bring the great bulk of its muscles into play, it does so anaerobically and runs up an oxygen debt that can take hours to pay off. Chased to exhaustion, a flatfish or a salmon may take a full day to recover, and that in well-aerated water. In a polluted, oxygen-deficient estuary, it may never make it.

Tuna are open-water animals, limited by their flat-out physiology to oxygen-rich seas. Given that, they score over more orthodox fishes both by having a huge gill area and by being warm-blooded, living faster and paying off their debts more rapidly than run-of-the-mill cold fish. A tuna can recover from exhausting activity in around two hours, a tenth of the time needed by a non-tuna, and only about twice as long as our air-breathing selves.

At this point it should perhaps be mentioned that there are tuna and tuna, and that not all of them manage their thermal physiology in quite the same way. Most of our knowledge comes from the smaller species, collectively referred to as skipjack, simply because these are possible laboratory animals, while the larger beasts are not. You can keep two or three kilograms of skipjack in a swimming pool. A five-hundred-kilogram, two- to three-meter bluefin, belting along at twenty lengths a second, is big for an oceanarium.

In skipjack, the muscular heat exchanger runs from the dorsal aorta and drains back into the posterior cardinal vein, holding the heat in strips of muscle on either side of the vertebral column. In the normal course of events, the temperature of this core is a mere two or three degrees above ambient. The situation changes when the fish become excited, dashing to and fro for food; the excess temperature then soars to two or three times the routine level. Skipjack, indeed, may be in some danger of overheating in a feeding frenzy, since they seem to have no means of bypassing the central core heat exchanger. Presumably, they carry on until they feel a bit strange, and then lay off for a bit. Lethal temperatures for tuna muscles are not known, but are unlikely to be higher than our own. The prognosis for a mammal at forty degrees Celsius is unpromising, and one must assume that the same applies to fish.

Thermal runaway becomes more and more likely as skipjack grow (the largest can attain ten or twenty kilograms)—the old surface-to-volume problem of all warm-blooded creatures (if an elephant had the metabolic rate of a mouse, it would catch fire, unable to get rid of the waste heat fast enough). Arguably, many features of the migrations and habits of skipjack are dominated by just this fact; the larger animals quit the tropics or move into deeper, cooler water, making forays to the surface to feed and hurrying back to the deeps to cool off.

The larger tuna species, and we are now talking of animals ten times as large as skipjack, with some real giants, weighing as much as three-quarters of a ton, operate heat exchangers located along the sides of the body, with strips of active, warm-blooded muscle fed from cutaneous arteries and veins. The dorsal aorta is reduced to a relatively insignificant vessel supplying the guts. These fish, unlike the much smaller skipjack, seem little affected by the temperature of the seas in which they choose to swim. The relatively superficial location of the hot bits apparently allows them to dump heat when they need to. At all events, the excess muscle temperature is greater at lower temperatures; the lateral muscles of a bluefin in water at 7 degrees Celsius can be as high as 26 degrees Celsius, while the same-size fish at 30 degrees Celsius has an excess temperature of only 2.5 degrees Celsius. A big fish, moreover, has considerable thermal inertia, so that an elevated temperature resulting from a period of vigorous activity can be made to last for hours. The net result is that large tuna can come very close to maintaining a high constant body temperature. They are, effectively, homeotherms. Certainly they are closer to that condition than are most reptiles, and over any twenty-four-hour period they are probably nearer to a stable temperature than are bats or hummingbirds. It gives them total freedom of the seas. Large tuna can and do move from the equator to the Arctic, swimming equally fast in both environments. They go back to the tropics to spawn, because the young ones, surface-to-volume ratio again, cannot maintain body heat in the coldest seas.

Some of the sharks have, apparently quite independently, evolved heat exchangers that allow them to maintain muscular temperatures well above that of the seas in which they swim. Mako and porbeagle have cutaneous arteries and veins like those of the big tuna species, and apparently maintain muscle temperatures well in excess of the ambient; freshly caught specimens are warm inside. The great white shark (*Jaws* prototype, and one of the oceanic sharks that eat people) has similar structures and is assumed to be warm blooded, though nobody has yet had the temerity to approach one in the sea and stick in a thermometer. One must assume that these fish, like the tuna, have warm guts and brains, too, and more heat exchangers, because a high-powered locomotor system must be supplied with fuel and a quick, decisive control system.

So much for the cold-fish story. Some, at least, are not. But why? Ask the question that is almost diagnostic of a zoologist, What is it *for*? By which the zoologist means, What was the advantage of this mechanism that led to variants with a tendency to warm-bloodedness being the survivors in a competitive world? The answer is not obvious; if it were, all or most fish would be warm-blooded, which most fish plainly are not. The penalty for being warm is a high fuel consumption. Skipjack (nobody knows about the sharks or the larger tuna) are somewhere between 25 and 40 percent more costly to run, per kilogram-kilometer, than nontuna at low speeds, a dreadful extravagance that should rapidly have put the brakes on any such evolutionary development. The high cost arises from a variety of causes. Tuna, being vertebrates, have a body salt content considerably less than that of the sea in which they swim. No fish can produce a urine stronger than the salts concentration of its blood; the extra salt must be excreted through the gills, which are also where most of the salt leaks in. A big gill area is necessary for the import of oxygen; a big gill area is bad because of the inward leakage of salt. No way of getting round this one. A tuna is estimated to expend something like 20 percent of its total fuel consumption exporting excess salt. A high temperature, whether of muscles or brains or guts, creates a high oxygen requirement, and that, in turn, requires a lot of blood circulating fast. Heartbeats cost an estimated 5 percent of the metabolic rate of a tuna. And so on. Tuna are high-cost fish, the racing machines in a underwater world where fuel economy generally has a high priority.

It must be worth it; the gamble evidently pays off. Tuna live mainly in the tropics, and although the more spectacular homeothermic species have now broken out of the necessity to live in warm waters, it is reasonable to assume that both the warm-blooded bony fish and the warm-blooded sharks originated there. Tropical seas are not rich in nourishment; anything that dies rots quickly and is rapidly recycled; there is no sign of the seasonal surge in productivity typical of temperate latitudes; plankton is thin, and midwater fish are scarce. Two strategies present themselves: move slowly and economically, play it cool, and subsist on the scraps of nourishment available. Some fish do this. The ultimate exponent is perhaps the sunfish, *Mola mola,* moving slowly and feeding, it is said, largely on jellyfish, themselves a dilute form of food. The alternative strategy is to gamble, flash from place to place extravagantly on the off chance of encountering one of the comparatively rare schools of fish, and make a killing when that happens. The tuna tactic. And, one must suppose, the tactic pursued by porbeagle and mako sharks. It is a good reason not to take any chances with a great white. It is probably bloody hungry.

DANGEROUS ANIMALS

South of latitude forty-two degrees we Europeans get into the zone where sharks bite people. North of that line, it is held, the water is generally too cold and the sharks too lethargic to bother. But even in the tropics remarkably few people are bitten by sharks. SCUBA divers, who spend a great deal of time swimming about in sharky places, hold that you should be wary rather than worried. My own experience of these matters is limited. I have frequently seen sharks while diving in Papua New Guinea and on the Australian Great Barrier Reef, and generally they have been swimming away from me, apparently having concluded that I was inedible. Night diving in such places is leery at first, but once you have fought down the uncomfortable feeling that there may be something large and hungry in the dark behind you, it is possible to convince yourself that sharks, being sensible creatures, don't want their food mixed with a lot of old iron and rubber, and you come to believe the statistical fact that a shark attack is rather less likely than a lightning strike.

Provided, that is, that you are not doing something stupid. Don't spear fish (you shouldn't anyway). Don't splash about like something wounded at the surface (a good description of my swimming technique), particularly in muddy water or in bad light. Don't set out to provoke a shark. And never touch them. Sharks have skin like sandpaper and a brush with a shark can mean abrasions and blood in the water, and blood in the water is asking for trouble. In general, stay out of the water if there are sharks about unless you have SCUBA and are prepared to swim underwater in an unfluttered manner.

So, forget sharks. The really troublesome animals are much less conspicuous. The odds are heavy that you *will* get stung by a jellyfish, sooner or later, and just a few of these can be dangerous. Avoid anything with an air float, on or below the surface. It belongs to a group called the siphonophores, and some of these pack a considerable punch. The Portuguese man-of-war, with a float that looks like a partially inflated plastic bag, has tentacles that trail for yards underwater, covered with stinging cells that can put you in hospital. But at least you can see it coming. The problematic animals are some of its relatively tiny relations, long, thin, transparent creatures that you will be hard put to see even with a face mask. In European waters, nasty stings nearly always come from small siphonophores, and you rarely if ever see what zapped you. There is very little that you can do to avoid them. If you get stung, call it a day and go swimming tomorrow—siphonophores have a patchy distribution. The stings will look inflamed, itch for a few days, but are not, in general, dangerous (there are always some people who react violently to any sting, sometimes because they have been stung before; if the reaction is bad, call a doctor).

In the tropics, things can be more serious. There is a particularly troublesome group of small jellyfish, the Cubomedusae (sea wasps), which can sting abominably (the welts from my own single experience itched for a month, and that was one of the comparatively harmless species). A few stings can kill you. Fortunately for us Europeans, the really bad ones live on the other side of the globe—Australians characteristically overdo everything, including sharks and jellyfish.

The occasional deadliness of the smaller jellyfish highlights a point of which any swimmer, or indeed a wanderer on land, should be aware. The animals that you have to be careful about are the small predators. Small animals that habitually feed on prey as large as or larger than themselves are particularly dangerous, because they are at risk from their victims. A damaged small predator is a goner. The courageous little creature has to go all out for a quick kill to avoid getting hurt in the struggle. So there has been tremendous selective pressure for the evolution of venoms of spectacular effectiveness. In this respect, some of the spiders, centipedes, snakes, and scorpions on land, and a number of marine creatures, including some of the jellyfish, are in an altogether different league from the bees and wasps that most of us consider quite bad enough. A bee sting isn't meant to kill you. The object of the exercise is to create educated neighbors who will leave bees alone and, if possible, instruct their offspring to do the same; kill off the intruder, and another idiot will move in.

The venoms of small animals are not, of course, aimed at people, or even, in most cases, at any sort of vertebrate. The common European octopus (which starts life as a tiny animal, for all that the ones you see are often quite large) produces a toxin that slays crabs within seconds, but it is crustacean specific and not much worse than a bee sting to us. The blue-ringed octopus (another Australian!), in contrast, produces a spittle containing a material called tetrodotoxin, which is one of the deadliest nerve poisons known. It works on practically anything, ourselves included—you die from respiratory block. (Exactly the same substance is found in the ovaries of pufferfish. The Japanese go to endless trouble to remove the ovaries and eat the rest of the fish, which everybody else very sensibly leaves alone.) The trouble is, you can never tell. A black widow or a stonefish is bad news for us, but it is inconceivable that the venom was developed for our benefit. We just sometimes happen to get caught in the cross fire from somebody else's war.

That being said, there are some elementary precautions that any curious observer of marine life can adopt to increase the odds against getting stung, or hurt in any other way. The most obvious of these are don't poke things with your fingers; don't put your hand into holes; look carefully before you touch even the most innocent-looking rock or patch of sand. And, don't mess with any animal that flaunts itself. If it is brightly colored or conspicuous and doesn't run away, there is probably a good reason for its confidence. Bear in mind that while we know some of the codes on land (for example, avoid conspicuous black-and-yellow insects, as most of them sting), we are often blissfully unaware of the rules underwater. If an animal displays at you, puffing itself up or posturing, back off. The odds are heavy that it is bluffing, but unless you are an unusually competent animal behaviorist and quite certain that you know what you are doing, or you recognize the species concerned and know that it is harmless, it is better not to take the risk. In the tropics, wear a shirt and trousers, and if you are going to poke about under stones or on the reef, put on gloves.

Most of these precautions are quite unnecessary in Europe. We live in an unusually benign environment, both on land and in the water. Nothing worse than a bee sting or nettle rash—we don't even have poison ivy on this side of the Atlantic. Between 1899 and 1965 there were twenty-six cases of shark attacks in the whole of the Mediterranean, fewer than one a year, despite the enormous numbers of people splashing about. Granted, thirteen of the victims died (sharks don't often attack, but they are good at it when they do), but the odds of obliteration in this spectacular manner are plainly enormously less than the chances of destruction by the propeller of a passing motorboat, or by a windsurfer, let alone by Latin drivers proving their masculinity on the streets.

Jellyfish, like mushrooms, are mostly quite harmless, and if you know how to recognize the few bad ones you can safely handle the rest. The trouble is, you don't. You should, to be completely safe, avoid contact with anything related to jellyfish. Sea anemones and the beautiful, featherlike hydroids that you can find even in rock pools are animals to treat with respect, because they all have stinging cells, like the jellyfish. Those of the vast majority are not powerful enough to penetrate the skin. The large, green and purple snakelocks that you find in rock pools all around the western Atlantic and Mediterranean seaboards will stick to a finger if you poke it, but the tiny harpoons with which it impales you don't, unless you have unusually sensitive skin, penetrate far enough to hurt. At worst, it feels as if you have been handling fiberglass. Some of the hydroids are more potent and will raise a rash if you brush against them.

If you are snorkeling with a mask, or diving with a lung, be wary of spiny fish. The scorpion fish that form so important a component of bouillabaisse are often conspicuous, and they don't bother to get out of the way, which should warn you. If you want one for supper, take a net, and don't touch the animal unless you can be quite certain of doing so without getting spiked. The punctures will hurt and go septic. More dangerous, because they lie in the sand with only their eyes showing, are weever fish, also good to eat, but also potentially something of a menace. The hollow dorsal spine, with a gland at the base, can inject a dose that will take you at least into the outpatient department. As usual, it is unclear whom this defense is meant to deter; the most likely bottom-feeding enemies are other, larger fish.

Beyond these, there is remarkably little that can do you any harm. None of the crustaceans will hurt you, if you keep out of the way of their nippers. Clams and snails are all harmless (but don't bet on this in the tropics; the beautiful cone shells are, again, small predators out for a quick kill, and one or two of these, with venoms to kill fish, can turn on you and have been known to cause death). Squid bite, but you will never manage to catch one of them underwater; they can see much better and move much faster than you can.

If you dive down and turn over stones, or roll over rocks on the seashore (put them back right way up, incidentally; you can murder a whole ecology by leaving a stone wrong way up), you will reveal a range of beautiful creatures to be seen and admired. The only ones remotely likely to fight back are a few of the largest walking worms, so-called polychaetes, because they are covered with little bristles called chaetae. The chaetae stick into your fingers, fiberglass again. Once in a long while the worm will lose its temper and flash out a pair of jaws, normally hidden inside the mouth. The nip is startling, but not dangerous. Fireworms, in the tropics, are polychaetes—bright orange or red and often prepared to expose themselves on, rather than hiding under, rocks. The behavior and the color should be warning enough. The fire is due to the chaetae, which are unusually sharp and numerous; they stick in your fingers and are so fine that you can't see to pick them out.

Again, the beast is not really dangerous; the object is deterrence, the creation of an educated public (fish, one must suppose, in this case) who let fireworms alone.

Much the same can be said of the conspicuous and often very plentiful sea cucumbers that lie about on the seabed, soft skinned and spineless. They rejoice in a number of rude names for obvious reasons. If you must pick one up to annoy your girlfriend, be aware that they have a rather disgusting habit of discharging long lengths of their insides through the anus. Their insides are white and sticky. It seems to be their only defense, and it won't harm you beyond being a pest to remove, but your girlfriend won't love you if she gets it in her hair.

Sea urchins are the subject of another article, already encountered in this book. They can be a bore because the spines break off if you stub your toe or sit on one. Each spine is covered in a thin layer of flesh, and that goes rotten. It is small comfort to know that the remains will come out more readily if you let them fester a bit, the suppuration loosens the spines in the skin. But as you've seen if you've read that chapter, there are a lot of things you should know about sea urchins.

Nobody has ever been hurt by a starfish, a seaweed, or indeed by most of the things you encounter on a seashore or swimming underwater. These things are not threatened by predators or large browsing animals, so they have no active defense. The beasts to watch for are those rather uncommon animals that have evolved to brazen it out, and the small, quick-kill merchants that survive on creatures as large as or larger than themselves. And one other category: the carriers of nasty diseases. Underwater this is not a problem.

ANIMALS ON ISLANDS

The trouble with a desert island is that it is, by definition, deserted. A wrecked seafarer is rather unlikely to meet a fellow human, let alone a member of the opposite sex. It is impossible to colonize an island all by yourself, unless you happen to be female and pregnant and have twins, one of each sex, or a male child who reaches puberty before you reach menopause. Taken all round, the odds are against.

So how is it that so many animals have successfully colonized every bit of land that isn't actually covered by the sea and, indeed, quite a few patches that from time to time are?

In 1883, the island of Krakatoa, in what is now Indonesia, blew up and sterilized itself. But even that unpromising patch of real estate was invaded practically as soon as it was cool enough to settle upon. A botanist, visiting the island three years after the eruption, which wrecked not only Krakatoa but also the neighboring islands some fifteen miles away, found eleven different varieties of fern doing well. The sizable animals took a little longer. A zoological expedition mounted twenty years after the event found a flourishing fauna that comprised not only birds and bats, perhaps not surprisingly, but also lizards, including a three-foot-long monitor.

Krakatoa isn't far from Sumatra, a matter of twenty-five miles. So a monitor lizard lucky enough not to attract sharks might swim it in a day or so, or turn up on a floating log. As islands go, Krakatoa isn't particularly remote.

But even so there is a problem. Krakatoa blew up and covered itself in hot ash. Sterile soil doesn't offer anything for animals until the plants become established. Only then can the herbivores move in, and only then, the carnivores.

Or so the story went.

Maybe the people who visited Krakatoa three years after the eruption were not quite quick enough off the mark to observe the first wave of colonists. In 1980 Mount St. Helens, in Washington State, exploded spectacularly and wiped out several square miles of vegetation in a fiery blast that flattened and incinerated a considerable area of the country north of the volcano—ash and pumice; no living thing survived.

This time there was a university practically on the premises, and right from the start scientists were able to monitor the recolonization of the new desert region. The first colonists were predators and scavengers. There was, surprisingly, plenty for them to eat, because the first thing that blows in, along with seeds and vegetable debris, is insects. A sticky-paper trap in such a region rapidly becomes plastered with the bodies of aerial plankton blown in with the wind. A jam jar filled with paraffin or antifreeze (St. Helens is high and cold) soon accumulates an astonishing quantity of insects that arrive apparently from nowhere. Tons to the square mile, and all good fat and protein for any opportunist who is fit enough to set about its more exhausted fellow travelers. The herbivores that arrive with the first wave have nothing to eat; the carnivores have a field day. Birds muscle in, as do spiders (most of these, too, are paratroopers, not foot soldiers, blown in on long threads of silk) and predatory beetles, all cashing in on a rain of food sailing in on the wind and landing—bad luck this time—on an ash field.

Only much later do the plants begin to grow in the soil formed from the bodies of the dead and the waste products of the living.

Much the same must have happened on any oceanic island emerging as a sandbank or created by some cataclysmic upwelling. While the rest of the world, somewhere over the horizon, is burgeoning with insects, there will always be a supply of food for predators and scavengers.

Birds, too, constitute a sort of aerial plankton. If an island is remote enough, the Galápagos, or Henderson, Ascension, or even Hawaii, the land-living birds will be accidental, storm blown, or hopelessly off course on a migratory flight. A founding pair or a flock may well find that the world they have blundered upon is not only flush with food, but also happily bereft of the predators that they had to avoid back home. The main problem may well be the danger of being blown out to sea again. Flight is both unnecessary and dangerous, and it is not altogether surprising to find that many of the inhabitants of remote islands are flightless members of families that are normally airborne elsewhere. The dodo of Mauritius was a pigeon of sorts, and a sitting duck for hungry sailors who were delighted to find fresh meat that, with no experience of predators, made no serious attempt to run away. The Latin name, *Didus ineptus,* says it all. The solitaire on Rodriguez went the same way. There is a charming flightless rail on Aldabra that would hardly have survived the move, happily abandoned, to make the island into an air base in the 1950s.

Any species that survives the first few dodgy generations soon overcrowds its environment. Competition between animals generally is overwhelmingly within rather than between species—we are not the only animal that is its own worst enemy. So the island population tends inevitably to set up pressures within itself that rapidly lead to specialization. One effect is a tendency to move to extremes; bigger may be more efficient, smaller may mean an ability to exploit patches of food too tiny to be worth seeking for a larger individual. Any individual able to crack a seed or poke into a hole just a little better than its colleagues has an advantage. This is always so, in any habitat, but the edge is more likely to be preserved in a limited island population, where a single variant constitutes a significant fraction of the stock and the enterprising individual is exploiting virgin territory, a resource so far uncropped by others.

The classic instance, which everybody quotes, is Darwin's finches. When young Charles called in on the Galápagos he was shrewd enough to notice that many of the terrestrial birds were closely related, despite a wide range of lifestyles, from seedeaters to insectivores and even a sort of woodpecker. A single colonizing species, blown over from the mainland hundreds of miles away, had apparently radiated to exploit most of the vacant niches that would already have been occupied by specialists back in the Americas. It was one of the bits of evidence that finally convinced Darwin that the *Origin of Species by Natural Selection* really was a tenable hypothesis.

Darwin had to go all the way to the Galápagos to see speciation at work because equally spectacular endemic faunas on islands nearer home had disappeared by the time people in Europe had begun to brood upon the possibility of wild animals' changing over time. A Neolithic Darwin could have made his observations a lot closer to home. Most of the Mediterranean islands carried their own species or varieties obviously derived from mainland relatives, and in some cases there is good evidence that these were doing well until man and his domestic animals arrived, as little as seven or eight thousand years ago.

Malta, Crete, and Cyprus all carried populations of dwarf elephants and hippopotamuses, as did many of the Greek islands. Neolithic people were getting ivory from Tilos as late as seven thousand years ago. Majorca had its own goatlike antelope, *Myotragus,* until as late as 2000 B.C. *Myotragus* did well, as endemics go, surviving for more than a thousand years after the arrival of the first human colonists, perhaps because the later settlers actually cultivated the animal, for once, instead of simply hunting it to extinction. There is evidence that they sawed its horns off, presumably to make it more tractable as a herd beast. It probably died out as a result of diseases imported with man's other herd beasts, diseases to which, as an endemic, it had no resistance.

Myotragus is a nice example of the sort of changes that can happen to animals on islands. Its eyes, unlike those of grazers almost everywhere, face forwards, a useful trait for an animal living on a rocky island—jumping about from crag to crag is safer if you have good binocular vision. But this is a luxury few ground-living herbivores can afford. They need eyes on the sides of their heads to provide all-round vision for detecting the stalking carnivore. Until man arrived on Majorca, *Myotragus* didn't need to worry.

The absence of predation allowed corresponding changes to the limbs of island species. Since it was no longer essential to be fleet of foot, island deer tended to become relatively stocky, more surefooted perhaps, and less liable to snap limbs, but certainly slower. Priorities change if there is nobody seeking to eat you. The larger mammal species typically became smaller than their counterparts on the mainland, where sheer size gave a measure of protection. In a situation in which emigration was impossible if food ran out, as it almost inevitably did as populations expanded in the absence of predators, a small individual stood a better chance of finding just enough food to get by. There is grisly evidence of mass deaths and bone defects that can only have been caused by chronic malnutrition at a number of sites. A history of repeated overbreeding followed by a population crash may have been typical of small island populations, further accelerating evolutionary processes, since the few survivors would, in effect, be recolonizing at each cycle.

Just how rapidly size changes can proceed in an island population has recently been demonstrated in a study of red deer on Jersey, Channel Islands. During most of the Pleistocene, ice-cap withdrawal of water resulted in sea levels substantially lower than they are now. Jersey was connected to mainland France and carried a population of red deer indistinguishable in size from those on the mainland. For a brief span of a mere ten thousand years, during the last interglacial period, the island was cut off, as it is now. By halfway through this period of isolation, the deer had shrunk to no more than one-sixth of the size of their relatives on the mainland, an almost incredible change in a little less than six thousand years.

Nobody knows what happened to the Jersey deer. Deposits from the end of the interglacial would be fascinating—did they go on getting smaller?—but may not exist, so we shall perhaps never know. At all events, the little deer vanished when the land mass was rejoined. They were, maybe, no match for a new wave of predators, or outgrazed by other species of invading herbivores, or simply unable to compete with their huge cousins when it came to the knock-down-drag-out fights which characterize the love life of rutting red deer. For once, it seems, it is unlikely that man was responsible; no human remains of similar date have been found on Jersey.

An island, one might note, doesn't have to be a patch of land. A lake is an island, and the shallow sea around a land island may be just as isolated as the land itself. Deep water is an effective barrier to many marine animals.

A few years ago I had a research student who regularly financed his not infrequent outbreaks of wanderlust by going to remote islands to photograph the fish. He would look up the geography, study submarine contours and ocean currents, and take off with SCUBA gear and an underwater camera to photograph species which, because he knew his stuff, regularly turned out to be rare or unknown, restricted to their own isolated patches of water. He paid for his expeditions by selling photographs to the publishers of coffee-table books and by bringing back live fish with his luggage, specimens to be flogged at great profit to aquarists who will pay large sums for the chance of showing something their colleagues cannot get. Ascension Island even issued a set of stamps to commemorate his visit; the three-penny stamp shows a beautiful little black fish with a yellow tail, *Stegastes lubbocki,* named after him.

Roger Lubbock was not only a man of great charm and persistence (he had to be; Ascension Island was a missile tracking station at this time, in the middle of the cold war), he was also a biologist of very considerable promise. The fish stuff was a hobby; his research was concerned with how animals and animal tissues recognize self, a matter of very great importance for a whole range of human problems, from transplants to cancer cells. He was killed in a car crash in Brazil in 1981; he was twenty-nine years old. Those whom the gods love die young.

DIVERSE DIVERS

After the war, in 1947, my parents took us to France, to the Mediterranean coast, where I swam for the first time with a mask and flippers. As a schoolboy I had read about marine animals, but I had never actually met any, because the coasts of Great Britain were barred throughout much of my school days. There were mines and sea defenses, and a schoolboy (so runs the military mind) might be signaling to submarines.

The south of France was a revelation. There were animals in the shallow water that I knew from textbooks, but had hardly imagined as being common. Sea anemones and sea squirts, snails and starfish, sea urchins and fish and hydroids, and even, miraculously, octopuses. Animals plastered all over the rocks, and, yet further, animals concealed beneath the stones that one could turn over. That holiday was a turning point; I think I determined to be a biologist then.

I realise, retrospectively, that I wasn't very adventurous. I swam around along the margins of the sea and I tried to spear mullet. I didn't have a spear gun, we were short of currency and not that rich anyway, and my bamboo pole with a nail lashed to the end of it didn't even seem to frighten the fish. I think I knocked one aside once in a while, but I never speared one. Retrospectively, I'm rather glad. I never tried diving to any depth, to see how far or how deep I could go. I wasn't that much of a swimmer, had always hated the cold swimming pools that school had thought likely to help form my character, and put up with the discomfort because the things to be seen in the water were so marvelous.

But the whole aquatic business had to wait. Within a couple of years I was called up, a conscript in the RAF, and although I volunteered for long-distance postings, they sent me to occupy Germany. There, I sat on a mountaintop, and learned to ski and ride ponies captured from the Cossacks. Later, as an undergraduate, I had forgotten about the sea, and as a postgraduate I found myself an entomologist, working for a Ph.D. in insect physiology.

At that point I was offered a job on the staff of the marine biology station, the Stazione Zoologica, at Naples.

So I quit being an entomologist, and exchanged the butterfly net for mask and flippers. The sea around Naples and its offshore islands, Ischia, Procida, and Capri, is full of fine things, but many of them shun the excessive sunlight and are best sought under the overhangs and deeper than I had hitherto dared dive. Besides, the underwater was littered with Roman remains. One could follow the flooded streets of Baia and collect potsherds from a drowned harbor at Gaiola, a bus ride from the lab where I worked. I got a lot better at diving during the first couple of years and could soon manage seventy feet on good days—rather terrifying at first, because I go negatively buoyant at around sixty, and thus had to swim up unassisted by the buoyancy that makes it all so easy closer to the surface.

I also learned to cheat my own physiology. If you want to spend more than about a minute underwater, it helps to overbreathe. Take a lot of deep breaths in quick succession. Overbreathing prolongs a subsequent dive because it removes carbon dioxide from the alveolar sump and thus also from the blood. Carbon dioxide in the blood combines with water to make carbonic acid, and acidity is what tells you that it is time to breathe again (it seems a funny way to organize the system, since acidity is only an indirect indication of oxygen lack, but don't blame me, I didn't design it). Knock down the acidity and you no longer feel that you *have* to breathe after the first thirty or forty seconds. After a good bout of overbreathing, I can hang on for a couple of minutes. (Actually, this is a good way of cheating on the breath test, because it also washes out the alcohol, which is what the little green crystals are after. Take a few quick, deep breaths as the officer with the bag approaches, and you may get away with it.)

Try it and time yourself. But don't use it, or at least don't use it excessively. That is the temptation in mask-and-flippers diving, and it is dangerous. The problem is that oxygen lack itself does *not* stimulate breathing or make you feel that it is now imperative to come up for another breath.

In physiology class practicals at university we learned this in a rather simple experiment. Breathe in and out of a bag, and the air rapidly becomes intolerable; you need desperately to breathe something else. Repeat the experiment with potassium hydroxide, which absorbs carbon dioxide, in the airway, and you will go on, quite happily rebreathing the same air until you pass out from lack of oxygen, at which point your partner disconnects you from the apparatus (you have to trust him or her) and you come to, a little dizzy and wondering what went wrong.

Underwater, you don't have a partner to tell you when to quit, so if you overdo it, you may never swim back to the surface, or you may land face down if you do. Water absorbs carbon dioxide very readily, so you may never feel the need to breathe again. Overbreathing is something to be very careful about.

Professional diving animals don't do it.

Seals breathe quite normally before a dive and they actually breathe out before going down. A good lungful of air prolongs the dive by adding to the oxygen you take down with you, and it helps us, but presumably not seals, psychologically. But it also adds to buoyancy, and that is undesirable because it increases the effort of swimming down, at least until the lungs are completely collapsed by the pressure (sixty feet in my case). So seals breathe out. And go down for a long time, sometimes ten or twenty minutes, some species for even longer.

Is one to assume that seals are just very tough, able to resist the categorical imperative of blood acidity telling them to resurface, to an extent unthinkable in people? No, of course not. Animals are not deliberately braver than we are; indeed, when bravery is tantamount to stupidity, they are less likely than we to push matters to the physiological limit. Seals have adapted to their situation by enhancing a series of reflexes, some of which occur even in people. For instance, when a seal dives, its heart slows. That happens in people, too, but only if they are sufficiently practiced at the diving business so that it no longer worries them; adrenaline stimulates the heartbeat and cancels out the reflex in most of us. And in seals (though not, in this case, in people), a series of reroutings of the blood takes place. Much of the blood supply to the muscles closes down. There isn't enough oxygen in the system to keep all the muscles going for ten or twenty minutes, and muscles, in the crunch, can run for a while without. The oxygenated blood is reserved for the heart and for the brain, running in short circuit to the places where it really matters. The muscles, meanwhile, run into oxygen debt, accumulating metabolites, notably lactic acid, that will be dispersed when oxygen becomes available again. On surfacing, the circulation changes again, drains all the crud out of the muscles, and the seal breathes heavily for a bit to convert it all back to acceptable, exhalable, wastes such as carbon dioxide, pays off the oxygen debt, and prepares for the next dive.

Very similar changes take place in diving birds. The heart slows; the circulation alters; an oxygen debt is accumulated in the muscles, where the acid metabolites cannot drive the "must breathe again" system until the blood reserved for the heart and brain itself becomes acid and devoid of oxygen.

The real experts, of course, are the whales, and the real experts among the whales are the sperm whales, which can dive for an hour or more at a time. They do it to catch squid. Details of quite how they do this are unclear. Sperm whales are a trifle large as laboratory animals. What we do know from harpoon-rope lengths, echo sounders, and the occasional drowned specimen that has tangled with an underwater net or cable) is that they go very deep, to a thousand feet or more. We know, too, that they have a lot of blood and that their muscles are loaded with myoglobin, another oxygen carrier. Myoglobin is the reason flesh is red; in sperm whales there is so much of it that it is almost black. So, the whales carry down a lot of oxygen for their size. They presumably do the seal/bird thing and reroute the blood. And they gain from sheer size. The metabolic rate of animals, the amount of oxygen consumed per kilogram per minute, falls with increasing size (see "How Old Is a Fish?"), so a large animal, even if it only carries reserves proportionate to its volume, inevitably wins out. For all that, a sperm whale is exhausted by the time it returns to the surface, and has to float for a while, paying off a massive oxygen debt after each long dive. A sitting target.

No whale ever gets the bends. The bends is a peculiarly human problem, brought about by the invention of diving equipment. It is caused by bubbles of gas coming out of solution in the blood and tissues. The problem, discovered in the early days of hard-hat diving, is that air, pumped down to a diver on the bottom, has to be at much greater than atmospheric pressure. At thirty feet, atmospheric pressure is doubled; at sixty it is trebled; at one hundred feet it is four times what it was at the surface. If the air doesn't reach you at this pressure, whether via a pump line or from a cylinder with a demand valve, you cannot inflate the lungs against the pressure of water around you. So you breathe, use up the oxygen. But the inert nitrogen in the air also goes into the lungs, under pressure, and dissolves in the blood, and thence slowly into all the other body tissues.

At the depths normally considered reasonable by sport divers, down to one hundred feet or so, no harm is done until you try to come up. As you rise and the pressure drops, the nitrogen in the blood and body tissues comes out of solution and expands, and that can kill you.

The bends is no joke. SCUBA divers, who have a demand valve that ensures that they are always getting air at the ambient hydrostatic pressure (four atmospheres at one hundred feet, and so on) have to know exactly how quickly the unwanted gas enters and disperses from the blood. It is a matter of survival. On the first dive of the day you can have one hour at sixty feet, and it is still safe to surface without decompression stops. Twenty minutes at one hundred. Five at one hundred and forty. Leave it any longer and you are pushing your luck. Divers are obsessive clock-watchers; they have to be.

It is, of course, possible to stay down for longer, and many professional divers do so routinely. But as soon as they are over the limit they can no longer return directly to the surface. They have to come up in stages, equilibrating and shedding gas at a series of decompression stops, which is fine, provided nothing goes wrong and they have plenty of air in reserve to fuel the stops. Or they can retire into some sort of chamber filled with compressed air, to wait while the pressure slowly falls to the point at which they can return to the "surface." It can be very boring. After a dive to two or three hundred feet a diver may have to decompress for a day or more, so that major projects are often carried out from permanent-high-pressure stations, and divers decompress once after working for a week or ten days. Pressure per se doesn't seem to be bad for people in the short term, but nobody is quite sure what happens if you do a lot of it. We are the first diving generation.

The apparently obvious solution, which is to avoid the bends by breathing pure oxygen, doesn't work. The trouble here is that pure oxygen, at a pressure of more than about two atmospheres, is toxic. It seems to be bad for the nervous system, and the lungs produce fluid. A combination of convulsions and pneumonia does not generate an encouraging prognosis, and divers with any sense of self-preservation avoid using the pure gas below about thirty feet. Oxygen kit was used by divers during the war—if you want to swim in and attach mines to an enemy vessel the use of oxygen avoids leaving a trail of bubbles; it is used by cavers, where the bulk of a full Aqualung is impractical; and it is no doubt useful for poaching salmon. The quick answer, for most of us, is don't touch it.

The dual problems of oxygen toxicity and the bends can be reduced by breathing a very weak oxygen-helium mixture. Helium escapes from the blood and tissues more readily than nitrogen, which reduces decompression times, and the proportion of oxygen can be adjusted to the depth so that the total oxygen pressure never exceeds the dangerous two atmospheres. But now we are getting into expense levels that exclude most amateurs.

And, anyway, who needs to go that deep? A mere voyeur will find all that any reasonable being could wish to look at in the first fifty or one hundred feet, and the expensive air that you have had to buy from the dive shop will last longer that way.

But it would be nice to go in without all that clobber. Sometimes I really wish I was a seal. Except that I can't stand eating fish with the scales on.

THE LIFE OF THE LUGWORM

My late uncle was a worms man. More accurately, he was a one-worm man, having made a not inconsiderable zoological reputation—he was a professor in London and a fellow of the Royal Society—largely from his work on *Arenicola marina,* the lugworm.

The lugworm is the animal that leaves worm casts all over the intertidal zone of beaches where the sand is mixed with a good deal of mud. It is quite large, as fat as your index finger and six inches long, and it is in many places prodigiously common, so that a worm cast every few inches indicates a great deal of animal hiding below the surface. Sea anglers use them for bait.

If you look carefully at the region around a lugworm cast you will generally find a small depression, a shallow, conical dip in the sand, a few inches away. The worm lives between the cast and the dip, in a semipermanent burrow extending straight down below the cast for maybe a foot before curving round into a horizontal gallery that comes to an end immediately underneath the sand cone. It is the business of the worm to eat sand. It does this industriously throughout its life, swallowing the muddy mixture; the dip is caused by the sand's falling in to replace the sand swallowed by the worm. The worm lives on the organic matter stuck to the surface of the sand grains. From time to time it stops eating and backs up the vertical part of the burrow to defecate, discharging the now comparatively clean sand grains onto the surface of the beach.

Uncle Gip became interested in lugworms in the 1930s, during a period when there was much discussion of the causes of animal behavior. There were then essentially two points of view. One was that animals were reactive machines, creatures that did things in response to external or internal stimuli. When a shadow passed overhead, they ducked; when they felt hungry, they sallied forth to seek food. Extending the matter to more and more complex animals, learning could be incorporated into this approach, beginning with conditioned reflexes—the dog salivates when it hears the dinner gong—and going on to quite elaborate trains of action—such as the effect of reading this article on your subsequent relations with lugworms. If one knew all about the instinctive responses of a species and all about the past history of the individual concerned, one could reliably predict exactly what it would do next.

The alternative view was that animals, far from being mere response machines, generated actions from within themselves. An animal could be expected to get up and go even in the absence of any change in the state of the world around it and regardless of whether it felt hungry, or thirsty, or lusty. Animals, according to this viewpoint, contain the springs of action within themselves. The "Chain of Reflexes" people said this view was tantamount to proposing a vital force for living things, a sort of mysticism, largely the result of introspection and not susceptible to proper scientific enquiry.

The response-machine lobby always had an escape. Even if an animal, placed in an apparently unchanging environment with all its immediate needs satisfied, did get up and go, one could argue that there is strictly no such thing as an unchanging environment (the moon goes round the earth, and who is to say that animals do not respond to gravitational changes, for a start) or that the animal had a history and was carrying on with some program of action set in motion long before the start of the experiment.

The lugworm, argued my uncle, was uniquely suited to a study of the driven from within or without problem. The range of external stimuli likely to affect a worm buried in the dark beneath a foot of sand was bound to be limited; it should be possible to identify all such stimuli and make a complete catalog of the events pushing the worm to carry out its rather simple actions.

In aquaria, in his laboratory, lugworms made burrows and continued apparently to behave just as they were believed to behave on the beach. They ate sand and made worm casts. Moreover, it turned out that they did this with remarkable regularity. Each worm would produce a cast every forty minutes or so. So far so good. It probably took that long to fill the gut with sand.

The bothersome thing was that they continued to show a forty-minute cycle even when placed in artificial burrows made of clean glass tubing with not a grain of sand in sight.

Uncle fixed up little floats at each end of the glass tubes, connected these to levers scratching a record on smoked drums, and left the worms to record what they were up to, hour after hour, day after day. It turned out, from these records, and from watching the worms at work in the tubes while the records were being made, that the full forty-minute cycle of activity went as follows.

Most of the time was spent lying in the horizontal gallery of the glass burrow. During this time the worm made movements as if feeding. It extruded the foremost part of the gut, which turned inside out as it emerged, exposing the inner surface. This is sticky with mucus, and in the normal course of events, grains of muddy sand would adhere, to be pulled inside when the pharynx was sucked back in again. Each extrusion and swallowing lasted about five seconds. After a number of these gulpings, the worm would stop and reflect for a while before resuming the endless task. A feeding-resting cycle would last six or seven minutes. Every forty minutes, after half a dozen feeding cycles, the worm would back up the vertical section of its artificial burrow as if to defecate, fail to do so—there was nothing in the gut—and crawl slowly back down again. During the return journey, the worm passed ripples of contraction forward along its back, drawing water down into the burrow through the tail shaft. In nature this would serve a dual function, irrigating the gills on the back of the worm and loosening up the choke of sand falling down at the head end of the burrow.

The records produced by worms in glass tubes were exactly like the records made from worms buried in sand, feeding and defecating normally. Quite obviously, neither swallowing nor defecation was dependent upon stimuli outside or inside the gut of the worm. *Oppure si muove.* The springs of action evidently lay within the animal.

At this stage, antivivisectionists should withdraw, because the next move was to search for the origin of the cycles within the worm, and that meant taking the worm apart. It turned out that there was no single origin. The fore part of the gut, removed from the worm and suspended in seawater, would go on rhythmically performing the act of extension and swallowing, every six or seven minutes, for hours on end. Rings of gut, or strips cut longitudinally from the esophagus, showed the same persistent rhythm. The more complex forty-minute defecation-irrigation cycle evidently depended upon pacemakers within the animal's nerve cord, but again there was no one site of origin; almost any bit of the cord could trigger muscular responses by the body wall at forty-minute intervals. The brain, at the front end of the worm, seemed to play no part in all this, and could be removed without visible effect.

So this simple animal, which, one would have thought, could quite easily have proved to be a response machine, in fact turned out to generate patterned activity within itself. It seemed very unlikely that these patterns were the result of learning, the more so since the limited brain that the lugworm possessed appeared to be irrelevant. And it would have been very odd to find the site of what was traditionally regarded as a higher nervous activity located in the throat. Here was evidence that even a very simple animal might include generators initiating actions that seemed to have nothing to do with external or internal stimuli. Each finding from *Arenicola* was a nail in the coffin containing the remains of the view that the activities of animals could ever be analyzed simply in terms of reflex responses.

We are now rather used to the concept of internal clocks in animals. Many marine creatures, for example, will continue to run their affairs in cycles that coincide with daily tides or the phases of the moon long after they have been removed from the sea and placed in aquaria. People, mice, and cockroaches all show persistent twenty-four-hour circadian (*circa diem*) rhythms if kept in apparently quite uniform surroundings. The cycles drift a bit over many days, itself an indication that they are not entrained to some invisible outside force, but remain remarkably closely locked to twenty-four hours—one reason jet lag throws us so badly. Attempts to locate the pacemakers show, as with the lugworm, that these are rarely well localized. More's the pity; if we could pin them down we could do something about it (well, you can always try melatonin).

Clocks geared to tidal, daily, lunar, even seasonal rhythms make obvious sense, as do pacemakers in organs such as hearts that have to remain continuously active throughout an animal's lifetime. But what is in it for the lugworm? Why shouldn't it be a reflex machine? On the face of it, it seems quite unnecessarily elaborate to equip an animal with a series of timing devices when a response to simple stimuli would apparently serve just as well. Animals are no more elaborate than they need to be. Natural selection does not preserve a complex mechanism when a simpler one will do; random changes to the genetic specification will degrade any system in the absence of associated selective advantages.

Consider, however, where the lugworm lives. Deep in the muddy sand between tidemarks conditions may remain quite uniform, but above, at the surface, potentially disastrous changes occur. The tide goes out, and puddles of water are left, which may be too hot and too salty from evaporation, or far too dilute for the worm's well-being if it rains. Conditions change irregularly and unpredictably. Down in the burrow, between tides, conditions are becoming unpleasantly anoxic. A worm responding simply to this situation would back up and draw down water of quite unsuitable salinity. If further wired to withdraw in this event, it would retreat down the burrow, only to be driven back again by a reflex imperative to oxygenate itself. Better, by far, that it possess an internal clock that will ensure on the one hand that it returns to sample conditions at the surface at intervals, in case things change for the better, while at the same time ensuring that it doesn't do this too often. Even a simple animal can benefit from a timepiece to program its actions and stir it into activity so that it doesn't slip into an irreversible lack of response as its sense organs adapt to slowly changing conditions.

BASKING SHARKS, ECONOMICS, AND POLITICS

Cetorhinus maximus is a big fish. Fully grown individuals can top seven tons and thirty feet in length. The part that you see in the distance, a familiar object to yachtsmen on the western seaboard and not all that uncommon in the Mediterranean, is the big triangular fin, archetypal shark, except that it seems rather floppy and slow moving to portray real menace.

The fish cruise along just below the surface on sunny days, mouths open, feeding on small planktonic organisms, which they filter through a huge sieve of gill rakers. Quite why they choose to proceed along at the surface is anybody's guess, because the plankton is not particularly plentiful there, and doing so renders the fish uniquely vulnerable to humans, who have been enthusiastically harpooning the poor brutes for centuries. As a species, they never seem to have learned. If you are sailing, or even motoring quietly, it is still quite possible to creep up to within prodding distance of a basking shark before it panics and dives.

Quite why basking sharks do anything is a bit of a puzzle. Despite their noteworthy size and their usefulness (their livers are enormous and full of oil, which is used in cosmetics and other specialized lubricants), remarkably little is known about the creatures. They appear, often alone, but sometimes in quite large shoals along the western Atlantic seaboard, starting off in Portugal early in the year and spreading up to the west of Scotland by May and June. Nobody knows whether the northward spread is a true migration or simply a reflection of the growth of plankton as the sea warms up after the winter dieback. It could be a series of populations that move in successively.

In autumn the sharks vanish. Presumably they retire to deeper water and sulk until the sieving is better again in the spring. Occasional captures—individuals that get tangled in nets set in deep water for other species of fish—show that these winter sharks have, remarkably, shed the gill-raker sieves, so that they are not even capable of feeding then. Maybe they just go and lie on the bottom, as some other sharks are now known to do (sharks don't have swim bladders and have to swim continuously to keep up; if they don't want to feed it is more economical to lie on the bottom). We simply do not know.

The list of things that we simply do not know about basking sharks includes almost everything about their life history, for a start. We don't even know how old they are. Other fish, cod and herring, plaice and such have scales, in which one can read the history of an individual, much as with growth rings on a tree. A fish's scale can tell you of its good and bad times, of its love life and its habits season by season. But sharks don't have scales. They have, instead, little toothlike projections, which give their skins a texture like unidirectional sandpaper (if you want to stroke a shark, do it head to tail, not vice versa, or you'll take the skin off your hand). The little teeth are useless for aging purposes. Nor do sharks have otoliths, the little lumps of calcareous deposit that bony fish have as part of their balancing apparatus. Again, this is a pity, because otoliths, like scales, have growth rings recording individual life history. It is difficult to age any shark, and almost impossible to say anything definite about basking sharks. Knowing how old the animals are is absolutely fundamental to any intelligent attempt to regulate a fishery or to conserve a threatened species, and for basking sharks (which might be a threatened species for all we know), we are stuck at square one.

Another set of vital information that is sadly lacking is anything that has to do with the reproductive rate of *Cetorhinus.* It is a strange fact that despite records of many thousands of sharks netted, harpooned, and processed, only one pregnant female has ever been caught. She was taken off Norway in 1936, and gave birth to five living young (and a stillborn) while being towed into harbor to be cut up. Each of the young was a meter or more long. This species, like a number of sharks, is viviparous. How fast the young grow is, again, anybody's guess. The guess is that they spend rather less than a year growing inside the female (otherwise a proportion of summer-caught females would be visibly pregnant) and that they are normally born during the period when the adults vanish in winter. But that is pure speculation. If pregnant females behaved differently from other females they could have a gestation period lasting for a year or more. Male sharks behave differently from females anyway, and maybe pregnant females go and join the menfolk. At all events, ten or more (nonpregnant) females are caught for every male basking shark, and in some fisheries a ratio of twenty or thirty to one is common. Why? It seems extraordinarily unlikely that any vertebrate runs to a sex ratio of ten to one (well, maybe turtles do, but that is a matter of the temperature on the beaches where the eggs develop, which wouldn't apply to sharks). Genetics simply doesn't work like that, and although some species of fish normally change sex in the course of their lifetimes (see "On Being Both Sexes"), there is no evidence that sharks ever do so. Leastways, we never catch intersexes. The only reasonable explanation is that males behave differently from females; they comb their plankton deeper and thus avoid capture. But why should the sexes behave differently? Males and females are of similar size, and they feed in the same way. They differ only in that males have a pair of pointed claspers (no penis in fish) formed from part of the pelvic fins. Dogfish have similar apparatus, and dogfish are known to use them to hook onto females while a packet of sperm is transferred. But, again, nobody has ever seen a basking shark mating.

What is needed is an extensive tagging program. A lightweight harpoon, fired to attach a numbered flag to the base of the dorsal fin, would do the trick. After that it would be a matter of alerting yachtsmen and fishermen—those who are not after basking sharks—to report sightings. (Commercial fishing boats often cheat on tagging returns because they realise that if they send back too high a proportion, the fisheries people will rumble that the population is being overexploited and will cut quotas.) If the number of tagged fish is known and the proportion of tagged animals is honestly reported, you have an estimate of the population, or at least an estimate of that part of the population silly enough to hang about at the surface where they can be seen.

There is evidence that tagging is unlikely to damage the sharks. In 1984 a single experiment was carried out in which a radio transmitter was attached to a basking shark. The transmitter was many times larger than a marker tag would have to be; in fact, it was carried in a little boat, half a meter long, full of batteries (progress—the same gear would be about ten centimeters long now). The boat was attached to the shark with a line to a dart through the base of the dorsal fin. Every time the shark surfaced, the transmitter signaled to a satellite, which reported the shark's position.

The shark was darted just north of the Isle of Arran, off the west of Scotland. It dived at once. It reappeared five days later, thirty miles south, close to the small island of Ailsa Craig, after which it just hung about for almost a fortnight, cruising back and forth. The feeding must have been good, or it may have liked the company of gannets—Ailsa Crag is one vast gannetry—but at all events it seemed to behave entirely normally until the transmitter broke loose and contact was lost. The experiment needs repeating. A series of such trials, together, perhaps, with a straightforward flag-tagging program, could tell us a lot about the movements and habits of the animals. But somebody has to pay for boat and satellite time. As usual, it is a question of who cares and what it is worth. The problem in this case is that shark fishing is a pretty small-scale operation so far as any one fishing boat or co-operative is concerned; nobody has enough capital in the business to give him an interest in the long-term conservation of the stock. And, as nearly always with a marine resource, nobody owns the animals anyway.

What is it worth? is, of course, a key question for people in my trade and for the animals we study. Zoologists are in an odd line of business, businesswise. If a zoologist gives the right advice about cropping levels, or crop protection, or conservation, the problem is reduced or goes away. He is continuously putting himself out of business. Worse, he is putting a whole lot of other people out of business. One could cite the use of insecticides on crops. Any manufacturer of insecticides has a vested interest in protection that is less than 100 percent effective. The last thing he wants to do is wipe out the pest or reduce it to such a level that next year, the client won't buy the treatment. If the pest develops resistance, so much the better. This means a bit more R & D, which means more work for the chemists and the biologists testing the goods, more work for the salesmen, who need to convert the client, and an excuse, perhaps, to introduce a product with a bigger profit margin. All good for trade and full employment, but insecticides often do a great deal of harm to the environment as well as some good to the crop being protected.

A good biological solution would be to change agricultural practice to favor natural predators or parasites, maybe import a new one or bring in an ally that somehow failed to arrive with an imported crop. There have been occasional spectacular successes with this approach, and some real disasters. Generally, it doesn't work well enough to be convincing (the ladybugs would have wiped out the aphids years ago if it was as simple as that), unless you have a public prepared to pay a little more for unsprayed foods, which is really only a possibility in affluent societies. The object should be not so much to eliminate as to minimize the need for chemical help. But who is interested in paying to do that? Governments are sensitive to anything that increases unemployment, even if it produces more food to feed the unemployed.

Commercial fishing, to come back to a marine example, is another case in point. The North Sea (among others) is overfished. Everybody agrees on that, and the fisheries biologists in the countries with boats in the North Sea know full well that the potential crop greatly exceeds the actual crop. They also know how to achieve it. Fish, like most animals, grow very rapidly in the period preceding their final attainment of sexual maturity. After that things slow down, until finally there is rather little growth from one year to the next. Since fish, by and large, lay masses more eggs than are necessary, the trick is to crop them as young adults, or just before they reach that state, but certainly not before they are well into the period of rapid growth.

Fish less and use wider-mesh nets. But that is not a commercially or politically acceptable solution. A trawler may have been built on the basis that it will bring in a 20 percent annual return on capital, but it is worth continuing to run if it brings in only 10 percent or 5 percent or (if you want to keep the crew on the payroll) 0 percent. Trawlers last longer than it takes to decimate a fish stock. So what government is going to allow the introduction of quotas that will oblige it to write off most of its fishing fleet, with the creation of a large pool of disaffected voters? Governments are introducing quotas, a bit at a time, because the scientific evidence of the depletion of fish stocks is irrefutable. But then they cheat pretty often at the fish-dock level by not noticing when the quota is exceeded. It is a brave inspector who really wants to risk getting thrown into the dock.

BUOYANCY

For most animals there is little point in floating about on the surface of the sea. You are dangerously exposed, you are silhouetted, which attracts the attention of subsurface predators, even supposing that the seagulls don't get you first, and there isn't a great deal to eat there unless you are particularly good at skimming bacterial films. Even that is hazardous now that suntan oil and supertankers have been inflicted on a once-clear sea. So only a handful of animals actually make their living floating on the surface. A few jellyfish, of which the Portuguese man-of-war is perhaps the best known, float with their tentacles dangling and dangerous in the water below and are left alone because they sting. And an odd insect, *Halobates,* a pond skater that has taken to the open ocean, survives because it moves so fast that nobody can catch it.

But a lot of animals spend their lives floating in midwater. They never touch bottom, and they never break the surface. It can be done by swimming for the whole of your life, and mackerel, for example, do just that. Mackerel are fast-moving predators, don't have swim bladders, and sink if they stop swimming. They have to keep moving anyway to ventilate their gills. Put a mackerel in a bucket of water and it suffocates, because it is normally ram ventilated, swimming with its mouth open to drive water in over the gills. In the course of mackerel evolution the swim bladder ceased to have a useful function; it merely added extra bulk to a fish, when sleek streamlining mattered above all things, and in time it disappeared. Most animals have been selected for less energetic lifestyles and economize by being neutral or nearly neutral in buoyancy, so that they hang in midwater with a minimum of effort.

There are several ways of adjusting the weight of the body so that it exactly matches the weight of the seawater that it displaces. The most familiar of these is the fish's swim bladder, a bag of gas under pressure, derived from the lungs of early swamp-living vertebrates. Fish with swim bladders are, geologically speaking, a comparatively recent invention. Fish manage by secreting oxygen from the blood into the bag, a remarkable achievement at any depth because the hydrostatic pressure pressing all round on the fish is all the time doing its damnedest to drive the gas back into solution. Fish succeed nevertheless in secreting gas against an uphill gradient by exploiting the properties of a countercurrent exchange system—a grandiose term for a simple bit of plumbing that turns up repeatedly in one form or another in circulatory systems throughout the animal kingdom.

Any plumber would know better than to run the hot and cold pipes to a radiator in contact with one another; the water moving in along the hot pipe would be cooled by the cold water coming out, and with any considerable pipe run the radiator would never heat up at all. Or, to put it another way, it is a good way to keep the heat in the boiler. Which is exactly what a number of animals need to do. Birds' feet get very cold; a seagull would lose a lot of body heat through the feet but for the fact that the arteries running down the legs closely parallel the veins coming back. Keeps the heat in the boiler, as do similar arrangements in the flippers of a dolphin, and, perhaps rather surprisingly, inside some parts of fish (see "Hot Fish").

What has all this got to do with buoyancy? Swim bladders in fish exploit a countercurrent exchange system to trap gas, not heat. The problem for a fish is how to build up gas pressure in a bag under considerable hydrostatic pressure, which tends to drive the gas back into solution in the blood, which has to be the source of the gas in the first place. The trick is to create local conditions that will cause the gas to come out of the blood and accumulate. Gas in blood is partly in solution and, more importantly, attached to blood pigments. Blood pigments give up their attached oxygen if acidified, so the first desirable local condition is acidity. A second condition is high salinity, because gases are relatively insoluble in very salty water. Both are achieved by salt pumps that actively transport salts out of the blood into the region immediately around the blood vessels bordering on the swim bladder. Active transport is fueled by oxygen, which is converted into carbon dioxide, which combines with water to make carbonic acid, so the mechanism producing a region of high salinity also acidifies the blood. Both the salinity and the acidity are trapped in the gas gland in the wall of the swim bladder by the same old system of parallel piping that a lunatic plumber would use to make sure that your radiator would fail to work. The gas comes off the blood, is trapped by the same plumbing that works for salinity and acidity, and has no place to go but into the swim bladder.

The mechanism is so effective that gas from the blood can be expelled against the pressure of several hundred feet of water.

Even compressed gas is a lot lighter than water, so a comparatively small swim bladder can balance the flesh and bone weight of quite a big fish.

The bad news for fish is that a pressurized bladder is a very inconvenient thing to have if you need to change depth. Dive down, and it compresses to a smaller volume and buoyancy is lost. Rise, and it expands. Fish can secrete in more gas or let gas off by expanding blood vessels—not part of the countercurrent system—in the walls of the bladder; the compressed gas readily goes into solution if given the chance. The problem is that both these mechanisms are rather slow, so that a fish making rapid changes in depth, anything greater than ten or twenty meters in an hour, is at risk. When a trawler hauls its nets, the fish come up too quickly to compensate and are killed by the expansion of their swim bladders. That, of course, is not the sort of situation that fish were evolved to handle, but it highlights a limitation on fish behavior set by their buoyancy mechanism. Fish with swim bladders tend to live horizontally rather than vertically, migrating maybe towards the surface to feed at night, but certainly not rushing up and down on a short-term basis.

Other animals have exploited safer, if bulkier, means of adjusting their buoyancy. Many small organisms retain fats and oils, which are lighter than water, to balance their body proteins, which are heavier. Some of the largest fish in the sea, basking sharks, which figure in their own chapter because they are remarkable animals in other ways as well, have huge, oily livers, while the marine mammals are covered in blubber for buoyancy as well as for insulation.

An alternative to fat is to fiddle with the ionic content of the body fluids. The commonest dodge is to trade sodium for ammonium. Sodium is a relatively heavy ion, seawater is sodium chloride; ammonium chloride is lighter, volume for volume. Ammonia is the end product of protein metabolism in most marine animals. Hang onto that and you have liftoff. Like the fats and oils mechanism, it is not limited by depth, and ammoniacal animals are found down to the deepest regions of the oceans. There is a price to be paid: greatly increased bulk. Ammonium chloride is lighter than sodium chloride, but only marginally lighter. So neutral buoyancy is achieved by this means only by retaining very large quantities of fluid. The situation is made worse by the fact that ammonium chloride is toxic. Delicate parts, such as nerves, won't work if the concentration is high. So the stuff must be isolated in special tissues, often pockets of spongy matter among the muscles, and this makes for some very weird and flabby animals. It also makes them taste disgusting. Many of the deeper-water oceanic squid fall into this category, and to our palates are practically inedible. Sperm whales love them, and, it is estimated, gobble down a greater tonnage of squid than the total human catch of marine fish. There is lots of protein still left in the sea, and it is only a matter of time before we, too, find a way of cropping and eating ammoniacal squid. The campaign to market squid fingers is doubtless in the minds of advertising executives even now.

Squid lack swim bladders, but their ancestors carried gas floats and some of their relatives still do. The white cuttlebone that you find washed up on beaches on the Atlantic seaboard of Europe and in the Pacific is the gas float from *Sepia,* the cuttlefish. It isn't a bone, but an internal shell that in the course of evolution has changed its function from protection to the provision of neutral, controllable buoyancy. Slice through it, lengthwise, and its porous structure is obvious. It consists, in fact, of a very large number of wafer-thin parallel chambers filled with gas, which in life would balance the flesh weight of the cuttlefish.

Examination of the state of affairs in living and freshly-dead cuttlefish shows that the chambers are not always or ever completely gas filled. The very latest chambers, at the front (blunt) end of the cuttlebone, are filled instead with a fluid almost indistinguishable from seawater. Work backwards, and succeeding chambers have less and less fluid in them. And it is less and less salty. That is the clue to the way the gas gets into the chambers. Remove the salts—active transport again—and the water follows, dropping the pressure in the chamber, so that gas bubbles in, out of solution in the blood, a beneficial form of the diver-threatening bends.

The neat thing about the way *Sepia* achieves neutral buoyancy is that there is no need to compress the gas, as there is with swim bladder. Provided the shell can withstand implosion, the gas will always be at atmospheric pressure, because the gas in the blood is in equilibrium with the gas in the seawater outside the animal, which is always at one atmosphere, in equilibrium with the air over the water surface. So *Sepia* can change depth without fear of the runaway expansion or collapse that threatens a fish. But *Sepia* doesn't take advantage of this capability because it is a bottom-dwelling animal, preferring to cruise along just over the surface of the sea bottom, collecting the shrimps and other invertebrates on which it feeds. It is handy for us, because we can bring the animal up without damaging it to study its behavior (which is fascinating) or to keep it alive and fit until we want to eat it (unlike ammoniacal squid, cuttlefish are excellent).

The ancestral cephalopod gas float was, it appears, external. It is represented in the fossil record by the countless shells of extinct forms, of which the ammonites are perhaps the best known. Like great coiled snails (and some of them were enormous, the size of cart wheels), these midwater and bottom-living animals dominated the seas for millions of years. The living creature, head, body, and tentacles, inhabited the final section of a shell otherwise sealed off into a series of gas-filled chambers. We are fortunate in that a similar creature remains, the nautilus. There are none to be found in the Atlantic, alas, but it is common enough on mantelpieces and in shell shops, six or nine inches across, smooth and white, with flame-colored markings radiating from the center of the spiral. Cut in half, lengthwise, the shell is divided into partitions connected by a tube that runs through to the body chamber. In life, a thread of tissue, complete with nerves and blood vessels, connects the buoyancy chambers to the rest of the animal. The mechanism is exactly the same as in *Sepia:* remove the salts, water follows salts, gas comes out of the blood. The interesting thing about *Nautilus* (the Latin name is, for once, the same as the common name), quite apart from its living fossil status (early nautiluses, not at all unlike those alive today, are found in the Palaeozoic), is that it lives in quite deep water, down to more than a thousand feet, on the steep, seaward slopes of coral barrier reefs. It can be trapped down there—the animals are scavengers and come to lobster pots—and hauled to the surface without harm. It is a deep-sea animal that will feed and fornicate within hours of capture, surviving a change in habitat that would slaughter any fish with a swim bladder (see "Life with a Living Fossil.")

There is always a snag, of course. In this case it is the bulk of the shell, a massively built tank, strong enough to resist the thirty-odd atmospheres of pressure that it will find at one thousand feet down. Almost 90 percent of the lift of the gas chambers goes into supporting the weight of the shell itself. So, once again, a safe, pressure-proof mechanism is achieved only at the cost of a considerable increase in bulk. Sad, that, because it is probably the reason the shelled cephalopods eventually died out. They couldn't keep up with, or, more probably, ahead of, the fish, which had hit upon a higher-risk but smaller-volume means of cutting their swimming costs. But that is another story, told in "The Dilemma of the Jet Set."

DRINKING SEAWATER

Our fishy ancestors evolved in freshwater; their ancestors came from the sea. The early fish inevitably had problems, because their body tissues were a lot more concentrated than freshwater and the stuff kept leaking inwards. Water is like that; it moves from less dilute to more concentrated solutions until the two even out, and the scaliest of skins cannot keep all of it out all of the time. Besides, the fishy ancestors had to eat, and it is difficult to swallow underwater without filling the gut with the stuff.

A fish with drinking problems could minimize the metabolic effort that went into bailing by dropping the concentration of its body fluids, thereby reducing the inward movement of water. Provided the body proteins could stand the more dilute internal environment—and one must suppose that this was a progressive adjustment, taking millions of years—a drop in bodily saltiness had the added advantage of reducing the struggle to acquire sufficient salts from a regrettably salt-deficient habitat.

By the time our ancestors had the enterprise to crawl out of the primeval swamp and see how the coal crop was coming along, they were already rather dilute animals compared with the overwhelming body of creatures, nearly all of whom have body salt concentrations approximating that of seawater. Our blood reflects this; it is only about one-third as concentrated as seawater. We are also noticeably intolerant of dessication, another characteristic of animals that have had a freshwater stage in their ancestry, with water the one commodity unlikely ever to be in short supply. Lose 3 percent of your body water and you are desperately thirsty. Lose 10 percent and you are a goner. We vertebrates are unusually sensitive. Animals that colonized the land without a freshwater stage, the great majority that got there directly, up and over the shoreline, are typically much more concentrated and typically far more resistant to drying up.

Our fishy ancestors originated in freshwater, and only then colonized the sea, no doubt to the dismay of a great many invertebrate animals that were doing very nicely without an invasion by landlubbers with jaws and predaceous dispositions. Perhaps rather surprisingly, the invaders did not revert to a seawater concentration of their body fluids; maybe they were already too specialized to return to the ancestral condition. At all events, fish, which had evolved under conditions in which salts were at a premium, suddenly found themselves in an environment where there was altogether too much salt, tending to draw water out of them, when previously their problem had been leakage inwards.

In freshwater, fish take up salts with their food and also by active transport through the gills. Chloride ions are trapped at the gill surface and passed in by ion pumps. Since they cannot possibly produce a urine that is more dilute than freshwater—which is near enough distilled water—bailing by fish necessitates a continuous salt uptake to compensate for the small amounts that they lose with every milliliter of their copious urine.

In the sea, the salt pump must be reversed, extruding ions that have got in with the food—most of which is as salty as seawater—and with seawater itself, swallowed alongside. No fish is able to produce a urine more concentrated than its body fluids, and if these are more dilute than the ambient, that is just too bad, the fish is now losing water, not salts, which it would have preferred to preserve. A very few fish, such as eels and salmon, which can migrate from freshwater to seawater and vice versa, have to rejig the system and reverse the ion pumps when they do so, a remarkable feat which always takes a few days during which they have to hang about in brackish water and be careful not to overdo the speed of transition.

The cartilaginous fish—the sharks and the rays—have achieved a partial solution to the salts and water problem by retaining urea, normally found only as a waste product, so as to achieve the same total body fluid concentration as seawater. That helps, because water now ceases to move out of the body, provided the body tissues can put up with all that urea, which of course they can, having been at it ever since the Palaeozoic. In fact, shark tissues are so well adapted to the stuff that they no longer work well without it. A heart from a freshly killed shark will stop beating if you put it into a bucket of seawater, and start again if you piddle into the bucket.

Reptiles do not, of course, have gills, and a turtle is faced with the same problem as a bony sea fish, without the fish's mechanism for dealing with the matter. Like fish and, indeed, like all vertebrates outside the mammals, turtles cannot produce a concentrated urine. Instead, they weep salt tears—active transport again—but this time through glands close to the eyes. Seabirds—

birds are descended from reptiles—enjoy a similar mechanism, so they, too, can drink seawater without concentrating the dilute body tissues that they have inherited from their forebears.

What has all this to do with us? A great deal, if you ever find yourself thirsty and at sea. Because the one thing that you quite certainly *cannot* do is drink the seawater. It is much more concentrated than your blood, and will only make things worse. Whales and seals and dolphins can drink seawater, but they have rather specialized kidneys capable of producing a urine that is very much more concentrated than you can manage.

So what can you possibly do about it? You cannot possibly avoid losing water, because you have to breathe. The air entering the lungs will be fairly humid in a boat at sea, but it will rarely be 100 percent saturated as it goes in, and it will always be 100 percent saturated when you breathe out. It will also probably be hotter, and hot air carries more water than cold. The water you breathe out leaves the salts behind it. So you lose water and become steadily more concentrated or quit breathing. Sweating carries out small amounts of salt, but certainly not enough to balance the water loss. So, you try to avoid sweating. In the early stages of incipient dehydration it is probably worth retaining your urine and drinking that, but you will soon hit a point at which the urine is as concentrated as the kidneys can manage, and after that there is no point in recycling the fluid.

Water is released from body tissues in starvation. Fats, in particular, yield some metabolic water as they are broken down, and I suppose it is possible that fat men and women survive shipwreck better than do lean ones (I have never seen an analysis of this) because they have more extensive water reserves, quite apart from the fact that they may be better insulated. Carbohydrates have nothing like the same gram-for-gram yield, and proteins help not at all because all the end products of protein metabolism are toxic and need to be washed away with water.

This is why life rafts should always include fishing tackle. When water is short, you can remain hydrated if you can catch fish and wring them out. Fish blood is as dilute as ours, and we can produce urine that is just sufficiently more concentrated than our body fluids to remain in balance if we drink that. Of course, you need to catch the right sort of fish. A shark won't do, because shark flesh and blood is as concentrated as the sea around it.

In fact, any vertebrate *other* than a shark will do nicely. A bird, say, or a turtle. Or a seal or a dolphin. Maybe the larger life rafts should carry a harpoon and a few hundred yards of line. Diving animals have lots of blood, and you could strain most of the protein off with the red cells, through a shirt.

Today's nasty thought.

THINGS THAT GO FLASH IN THE NIGHT

Sailing at night in seas that luminesce is something splendid that is not given to all men. On a quiet night, with just enough wind to ghost along without the engine, it can be euphoric. Euphoria is worth seeking; we don't often achieve it in this rush-around world. You need a pause, or you miss it. Sitting in the cockpit on a night watch, it takes a little time to become adapted to the dark and to realize how much is going on in the sea around you. What you generally see first is a cloudy luminescence. Intermittent larger flashes add to the background, stirred by the bow wave and the turbulence behind the boat. It is better, if rather reprehensible, to switch off the navigation lights. Rarely, something big ploughs through, creating a wake of its own. Dolphins look good at night and are recognizable because they snort. Even more exciting are presences to which you cannot put a name, fish, perhaps, raiding the surface plankton, or squid. One can always hope for a sea serpent.

Or a mermaid. Years ago, we used to go to the marine biology station at Naples (I had a job there, once). Naples is a crowded city, full of rugged individualists, each doing his own noisy thing and all pressed together so tightly that you have to be a quite remarkably well-adapted human being not to scream, once in a while. Neapolitans, who are better-adapted humans than most of us, wouldn't understand and would certainly be offended by the screaming. So misanthropic northerners need an escape. Short of transcendental meditation, a reaction that would puzzle the natives, the best answer is a boat. We kept a small motorboat in the harbor at Mergellina, and in the evenings, as we finished work and darkness fell, we would go swimming (offshore it is, or was, pretty clean in the Bay of Naples; it is just along the shore that a thin film of suntan oil and plastic bags makes it look disgusting). The trick from the old hands who had been to Naples before was to persuade the more nubile of our student assistants that it was quite unnecessary to wear bathing costumes in the pitchy darkness, and some of them never noticed the glow around them as they swam. In the late '50s, you understand, people wore clothes at sea, even in the Mediterranean.

The creature that framed the girls so successfully was almost certainly an alga, or, rather, many thousands of algae—plants, not animals, unicellular and planktonic. The largest of these beneficent organisms, called *Noctiluca* (for once, a Latin name makes obvious sense), is almost a millimeter across and practically transparent, so that it is difficult to see even if you know what you are looking for. If you switch on the light you cannot see the flash. The best bet, if you want to examine one more closely, is to pump water into the boat's loo, in the dark, and make a grab with a glass while the poor creature is still upset and glowing. The mate will complain about the glass, and you must plead the progress of science.

Anyway, it is not much to look at if you haven't got a microscope aboard—most yachts are hopelessly underequipped—and a more interesting question is not what *Noctiluca* looks like, flashing or quiescent, but why it bothers to do it. Isn't it suicidal to advertise your presence in this blatant way, when you are near enough invisible if you sit tight and do nothing? The answer, for the alga, is copepods, the next step up in the food chain and the most abundant small predators in the sea. Crustaceans, they are related to shrimps, crabs, and lobsters. But they are only three or four millimeters long—big enough to see in a bucket, if you shine a bright light and watch for the shadows, for they, too, do their best to be transparent. Copepods move in jerks, pouncing on yet smaller water fleas.

They don't like prey that flashes. Grab it, it explodes into light in their scratchy little arms, and they drop it. The alga, flagellated as many marine algae are, paddles off. An odd bit of behavior, really, because the alga, so far as we know, doesn't taste nasty, or sting, or do anything else unpleasant to the predator (is a creature that eats plants that swim about a herbivore or a carnivore?), and you might reasonably expect the copepod to habituate to the flashes after a few tries. But it does not, apparently, and since practically every animal that has ever been studied habituates to stimuli that repeatedly prove harmless, there must be some other explanation. The most likely possibility is that the copepod itself wishes to remain invisible. The chemicals that provide the luminescence are quite simple, a matter to be considered later, and the trick is not so much to induce luminescence once you have synthesized the ingredients, as to keep the reagents apart when it isn't wanted. Chomp down the alga, and the unfortunate copepod would be, for a while, lit up by the glow in its gut, a sitting target for a fish or anyone else with a taste for miniscampi.

Whether the same applies to the larger flashers, we don't know. Most of the bigger burn-offs that you can see from a boat on a dark night come from ctenophores or siphonophores. Ctenophores are themselves predatory, sea gooseberries a centimeter or two long and hardly worth anybody's eating, largely composed of jelly rather thinly spread over with flesh. And siphonophores sting, quite nastily (see "Dangerous Animals"), so that no fish in its right mind would touch them anyway. Yet the flashing is plainly defensive; they only do it when disturbed by touch or turbulence. These animals do not copulate or fight one another, so there is no parallel with glowworms or other fireflies, or fish, or squid, some of which certainly do use flashes to communicate with one another in love or hate. As so often with marine organisms, we can readily observe a phenomenon, but what it is *for* is another and altogether more difficult and interesting matter.

Luminescence in animals that use it can happen in two ways: do it yourself, or cultivate luminescent bacteria. Some bacteria, and a few fungi, glow quite brightly in the dark, as you are liable to discover if your housekeeping is less efficient than it should be—meat and fish can be quite pretty in the dark if they have been around for a while. In the days when people used to cure their own hams, as my grandmother did, there could be an appreciable glow in the larders of quite well-regulated households. Times have changed with refrigeration, and anyway, you would miss it, because refrigerators light up when you open the door. I don't suppose that even morgues glow as much as they used to now that fluorescent lighting is so cheap to operate; brains were said to be particularly prone to glow, with those of dead maniacs, it is reported, the brightest of all.

In the sea there are some half-dozen sorts of luminescent bacteria, and at least three of these live in or on fish and/or with squid (among others, with some species of *Sepiola*), apparently to their mutual benefit. The bacteria cultivators are nearly all deep-sea animals, which makes their study a bit difficult, but it seems that in all cases the host fish or squid provides pockets in which the bacteria lodge and multiply, feeding on mucousy material secreted by the host. The whole affair is evidently finely regulated, because the bacterial culture must somehow be kept growing, dead and excess individuals must be discarded, and somehow the bugs must be kept from escaping to spread outside the pit or pouch where the light production goes on. A major problem about bacterial luminescence is that once set in motion by breeding up a culture of bacteria, it won't stop. The light is on all the time, so that if the fish or squid wishes to shut it off, it must have some sort of movable screen to shield the glow. Solutions range from a muscular shutter, to rotating the whole photophore so that it shines inwards against a solid wall of flesh.

Perpetual light is not a problem for the do-it-yourself school. That includes the algae and jellyfish mentioned already, through a range of other invertebrates, and again into fish. Here the name of the game is the production of a substance, luciferin, that can be oxidized by an enzyme, luciferase. Luciferin and luciferase are generic, descriptive terms for whole classes of materials that happen to act in this manner. The sticks that yachtsmen and divers buy as emergency lighting—bend one and it glows for half an hour or so—are based on the same principle. Cracking the stick releases an oxidant, generally hydrogen peroxide, that burns off the luciferin. In animals, the release of the activating agent is generally controlled nervously.

The capacity to synthesize luciferin and luciferase has apparently cropped up several times in the course of animal evolution, and the detailed chemistry differs somewhat from one group to the next. Some animals cannot make a luciferin for themselves and must pick it up by eating animals that can—the production of an illuminant is not an unmitigated blessing; what makes you repulsive to one predator may be exactly the reason another is hunting you down.

But again, the interesting question is, What is it all *for*? Why does a fish, or a squid, or anybody else for that matter, go to all that trouble? Many of these animals are quite large, and those that eat them are larger still, so that the luminescent-gut explanation won't hold; the animals are not themselves transparent.

There is no single answer. In some species, in which the patterns of lights differ between sexes, it is a fair bet that we are looking at identification signals, a means of bringing the sexes together in places that have so little light that shapes and normal, reflected colors won't do. Flashes can keep a shoal together, or warn a rival off a patch that the owner wishes to crop or defend. Others use lights to attract prey, from deep-sea anglerfish with luminous tips to rods that wave seductively in front of cavernous mouths, invisible in the abyssal dark. Or searchlights. One outstandingly devious genus of deep-sea predatory fish has developed a system of red searchlights that are probably invisible to everybody except themselves. Red light hardly penetrates beyond the first few meters at the sea surface, and most mid-to-deep-water marine animals are red-blind. *Malacosteus,* one must suppose, has the jump on almost everybody around: night sights to target prey that will never even be aware that they have been spotted.

But by far the most widespread use is camouflage.

Deep down, or even in shallow water at night, an animal is often invisible except as a silhouette against the feeble light filtering from above. A predator can come up from below, itself invisible to a victim in the water above it. Lights along the underside, carefully matched to the light coming from above, can destroy the silhouette. In principle, the system would work in broad daylight. Indeed, extensive experiments were conducted between wars to hide military aircraft in just this manner. The snag was the enormous power required to match sunlight, and for a long time, even military aircraft did not have that sort of power to spare. The system was used briefly at the beginning of World War II, for aircraft hunting U-boats. It reduced the range at which they could be spotted from a dozen miles to around two, uncomfortably close for a submarine recharging its batteries at the surface and needing time to crash-dive. But then radar became commonplace, and lights became obsolete.

The intensity required precludes the use of lights by shallow-water fish during daylight, but the situation changes at night or at depths where the ambient light is down to nighttime values even during the day. Careful measurement of the light output of light-emitting fish and some other organisms shows that many of them glow below with an intensity that accurately matches the likely glow from above. In at least some cases, the light-emitting photophores include a pair that can be seen by the eyes of the lightmaker, so that the fish (it appears that only fish and squid are quite this sophisticated) can fine-tune its own light output by regulating the release of luciferase, or pulling down the blinds, to match the intensity of the light currently filtering down from above.

But these are not the animals that you see from the deck at night, looking down from above. Silhouette blotting is a game of stealth, played out in the deep waters in an environment as foreign to most of us as is outer space, and just about as expensive to investigate. Once upon a time I had dinner with the captain of a hunter-killer submarine, a man who knew far more about the sounds that animals make underwater than I, poor ignorant zoologist, shall ever know. His ship had a reference library in its computer that could match up any bump, burp, or cackle with species that ranged from whales to snapping shrimp. I was so impressed by this that I forgot to ask about lights. No doubt he had an image intensifier and an identification kit for all those as well. I wonder what he did for luciferin?

DOLPHINS

Once upon a time, there were no mammals in the sea. A Mesozoic yachtsman would nevertheless still have found dolphins. The Mesozoic ichthyosaur was a reptile, a dolphin replicate, with a longer snout and bigger eyes and a tail that extended up and down rather than sideways, but recognizable, for all that, as a dolphin. We like to think that they would have had the same plunging habit—they breathed air—and the same healthy irreverence for things (like boats and plesiosaurs) that were plainly less maneuverable than themselves. It is nice to think of ichthyosaurs playing in the bow wave as mammalian dolphins do today. One of my dreams is that I am looking over the bow and see an ichthyosaur. But the dream has a bad ending: by the time I have got the camera, it has gone—the fate of serpent spotters everywhere. It is also, sadly, probably a nonsense dream, like my dream of living to play squash at a hundred and ten to the horror of the less-fit young. Hard-headed animal physiologists and anatomists point out that ichthyosaurs had vertebral columns quite unequal to the thrust of a dolphinlike tail sweep and arms that implied a rowing motion; they probably paddled around gently, chomping ammonites that were themselves unable to swim swiftly away. So much for romance.

The fact of history is that the ichthyosaurs were so good at coping with the Mesozoic ecology that nothing much happened in the way of mammalian sea fish until the good Lord saw fit to wipe out most of the reptiles at the end of the Mesozoic. A typically irresponsible act. Of all the magnificent reptilian fauna, the disappearance of these splendidly adapted creatures was the most incomprehensible. We can cook up all sorts of plausible excuses for the departure of the terrestrial dinosaurs, ranging from the dust-cloud winter following a massive meteoric impact, through a volcanic holocaust, to constipation when the flowering plants replaced the ferns. On land, the environment is forever unreliable. But what went wrong in the sea? Two-thirds of all the known marine species disappeared at the end of the Cretaceous for reasons that are still obscure. A drastic reduction in the extent of the continental shelf, caused by an accumulation of ice at the Poles, maybe, a volcanic holocaust, the impact of a meteorite in Mexico . . . whatever the cause—and it was a crying shame about the ichthyosaurs—the catastrophic demise of almost everybody who was anybody opened up the way for a whole new sequence of sea creatures, recolonizing and refilling the vacant ecological niches. It was a gold rush, in which the mammals, some of which had miraculously survived the catastrophe, rapidly laid claims and made good.

The great whales and the mammalian dolphins are geologically recent. While the reptiles, in the first glorious exploitation of the land by vertebrates, were still around, there was no possible opening for the mammals. Reptiles did everything better. They flew. They swam. And they ran about, many of them swift and bipedal, grabbing at the good things in their world with such ferocious efficiency that the poor little mammals never got a claw in edgeways.

The story that the mammals eventually beat the dinosaurs because the dinosaurs were thick (all that bit about the brain in the small of the back) is hogwash. Just watch any lizard, hunting along the top of a wall or among the herbage, and try to imagine what it would have been like to have a good-sized bipedal reptile on your tail. The notion that you only had to wait for nightfall to get the jump on your cooling predator is probably also nonsense. Quite apart from some, really quite convincing, evidence (predator-prey ratios, marrowbone structure, and so on) that the dinosaurs were actually warm-blooded, the very size of the larger ones meant that their problems were related to the dissipation of heat rather than otherwise. In hot weather they may actually have been happier at night than during the day. No, the fact is that it was a noncontest. Our ancestors were forced into the undergrowth, little, crepuscular, burrowing things that fed on insects, waiting for the ill wind that was due to blow some good on somebody. They waited for a hundred million years.

The early history of what happened next is so far undiscovered. Large skeletons of creatures that one might think of as seals or whales turn up in the Eocene, and we assume that these gave rise both to the toothed cetaceans and to the filter-feeding giants of today, far bigger creatures, incidentally, than the world ever knew in the time of the dinosaurs. That in itself is an astonishing fact; with a hundred million years in which to evolve, the absence of really enormous marine reptiles is extraordinary, an observation that should surely tell us something about the marine ecology of the Mesozoic.

We know rather little about the lives and times of the great whales. A whale with an explosive dart in its lungs is unlikely to reveal much about its normal behavior, and until really very recently that was about the only sort of contact that mankind was seeking to make. Dolphins have been less methodically molested. We can keep them in zoos, and, since the era of SCUBA diving, we have been able to see a few of them at work in their normal habitat. One of the extraordinary things about dolphins, moreover, is that occasional individuals seem positively to seek out human company, so that a degree of mutually investigative interaction is possible.

Thank goodness for dolphin curiosity. They brighten the place up. The sea, with due reverence, can be boring, and the boisterous enthusiasm of dolphins is a tonic for the yachtsman fretting over the state of the tide or the weather forecast. Dolphins plainly consider the world a splendid place, full of variety and happily stiff with fish.

People who write books about dolphins hold that their company is therapeutic and cite instances when a psychological laying on of flippers has brightened the lives of more than one gloomy introspective. I believe them. The manifest good nature of dolphins, who have every reason to be leery of people (we swipe their fish, drown them in our nets, and sometimes even shoot or spear them), is infectious, like the—also therapeutic—bounce of a puppy or a kitten. One can only hope that these occasional encounters do as much good for the dolphins concerned, because a dolphin that seeks human company must be a scientist, or some other sort of misfit, who cannot quite make it with his own species; *his,* because these solitary individuals nearly always seem to be males, and there must be some message in that.

So, apart from the fact that some of them are crazy, what is so special about dolphins? Well, for a start, the story is that they are quite intelligent. Intelligence is a curiously indefinable asset. We have difficulty rating it even amongst ourselves, so much so that practically no two experimental psychologists, educators, or sociologists can agree on suitable tests. A major block is cultural. Your perception of the way the world is and your capacity to score in intelligence tests plainly depends on your upbringing. A European would fail miserably in an intelligence test devised by an Aborigine. So we have very little chance of making an objective measurement of intelligence in any other animal. The best we can do is to observe that the animal seems to assess the likely outcome of complex situations by putting together information from several sources. Usually that means noting that the animal learns by experience in circumstances in which we would learn ourselves, an anthropocentric view that assumes that we have the answers and that other animals are struggling to catch up.

Dolphins, even by this rather dubious criterion, are intelligent. They learn fast in captivity. They can be trained to discriminate, recognise hand signals, even develop a sort of syntax for auditory communication with their captors. They may not have the nous to have learned to avoid people but they *have*—it is repeatedly pointed out—got very big brains.

The credit given on account of this last factor shows how fragile human intelligence can be. The reasoning applied runs somewhat as follows: Animals with big brains seem to indulge in more complex behavior than do pinheads; the brain is therefore the seat of intellect. Dolphins have big brains, so they are intelligent, sensitive fellow creatures. Save the whale.

The trouble with the argument from big brains is that it doesn't even work for man. Throughout a large part of the last century scientists did a great deal of head measuring in an attempt to assess the relative status of people, within societies and between races. A lot of dangerous conclusions were drawn as a result. White Europeans and white Americans came out on top, with the widest hatbands and the greatest cranial capacities. The lesser breeds of mankind, blacks and Native Americans, aborigines, and women of whatever race, had smaller brains and were plainly in their then-inferior positions for good reasons. Craniology became a major science, confirming what the white, Anglo-Saxon male already knew to be true.

The flaw, in the human instance, was that craniologists tended to be a little vague about scale and age effects. Women are generally smaller than men, and brains shrink as people get older, so that samples are liable to be skewed in ways that make them difficult to compare. Attempts to show, for example, that white-collar workers had bigger brains than laborers foundered in statistics—human variability is too great. And there was the embarrassing evidence of Neanderthal man, who had a larger brain than any of us and nevertheless somehow contrived to get himself extinct.

The brainy dolphin argument is flawed anyway, because the two sorts of brain, man's and the dolphin's, have evolved to do quite different things. Dolphins seek their prey in water that is often cloudy with plankton, turbulent and muddy, or just plain dark. In these circumstances vision is a poor way to find fish, and we must suppose that the early proto-dolphins, lacking the pressure-sensitive lateral line of fish, were driven to discover their whereabouts much as a blind man does today—make a noise and listen for the echo. If the system was crude to begin with, it is anything but crude now. Dolphins squeak underwater and listen to the echoes. The analogy with bats is obvious; they, too, squeak and listen. But the similarity practically ends there. Bats are, forgive me, batting on a far easier wicket. Air is a lot less dense than water, and flesh is mostly water. The head of a dolphin is of much the same density as the sea in which the animal is swimming. So there is little or no reflection; the sound waves tend to pass straight in and out the other side. That makes noises difficult to hear and sound sources very difficult to locate; the skull does nothing to blanket noises from one side, and both ears get an equal signal. Lie in the bath with your ears under the water and turn on one of the taps; it is almost impossible to determine which one is dripping by sound alone. A SCUBA diver can hear boats miles off—sound travels four times as fast in water as in air and attenuates only very slowly—but trying to sort out precisely where they are and whether it is safe to surface is another matter altogether. Dolphins and the other whales do a lot better than we do because their auditory apparatus is largely insulated from the skull. Their inner ears are enclosed in dense, bony boxes, foam-sprung to avoid direct transmission through the flesh and bone of the head, linked to the outside by wax plugs so that the receiving microphones are almost as directional as our own ears.

Dolphins, underwater, emit high-pitched squeaks, generated by driving air through a reed pipe in the nasal passages. The skull, an odd shape concealed beneath the fleshy dome of the forehead, forms a reflecting dish, beaming the sound forward.

Interpreting the echoes is the difficult bit. A boat's echo sounder emits bleeps of regular duration at regular intervals, times the return of the echoes, and, if all goes well, tells you how far away the bottom is, if weed, fish, garbage, and an excessive degree of heel don't interfere.

But supposing you were after more information than a simple bounce off the bottom can give you? Throw the same signal horizontally, and a whole range of echoes of varying intensities come back. Echo quality depends on the difference in density of the reflector and the medium. A swim bladder in a fish will give a good echo, because air is less dense than water. It will give an even better echo if the wavelength of the sound signal approximates to the size of the bladder, hence the high frequencies of most of the sounds that dolphins make. Middle C, with a wavelength of five and a half meters underwater, would do fine for locating a submarine, but is not so hot for finding herrings. High frequency equals shorter wavelength equals better resolution, but it increases attenuation, and decreases the range. Echo intensity drops as the distance squared. A large object at a distance can yield an echo of similar intensity to that of a small object close by. A squid doesn't have a swim bladder; rock reflects well; weed varies as it stirs in the current. Everything, living or dead, reflects an echo that will change in a characteristic manner with distance, orientation, and relative motion. Add to this the fact that many animals make sounds of their own, so that echoes have to be distinguished from primary sources. The problem of analysis moves from the merely horrendous to the diabolical even before the dolphin brain has to start computing an appropriate course of action.

So, dolphins have big brains. That doesn't necessarily denote intelligence, just the appalling difficulties of signal analysis in the modality in which dolphins are obliged to work. A memory man, who can remember who played outside-right for Sheffield Wednesday at every match for the last fifty years, is not necessarily intelligent. The memory man may not have a lot of time left for anything so imprecise as integrated thought. The dolphin's aquatic memory bank may require an awesome number of neurons to operate an identification system that would be the envy of any submarine commander, but a library is not intelligent.

Save the whale, by all means. But not on the rather dubious evidence of brain size. There are lots of better reasons.

FOOD FROM THE SEA, VEGETARIANISM, AND THE RIGHTS OF ANIMALS

A cruising man should know how to supplement his rations with fresh food. At the very least this means that he should tow a line for mackerel and the marvelous small tunny and dolphin-fish that one can collect farther south whenever the sailing is a bit slack and the boat is going slowly enough (somewhere between five and six knots, in our experience). But he should be more enterprising than just that. Shellfish generally are plentiful, for free, provided you are wary of oyster beds, and very rarely cripple you with a resentful gut. The European community has complicated regulations about marketing shellfish when the e-coli count exceeds two hundred and thirty bacteria per hundred grams, a tiny number, to be on the ultra-safe side. I suspect the regulation is, or more tactfully let us say was, widely ignored. Some of the best mussels I ever ate were grown at Banyuls, in the south of France, close to where we keep our boat and within a quarter of a mile of the town's largely untreated sewage outfall.

Mussels are good almost anywhere. The ones you glean off the rocks won't be as big as those grown on ropes around continental harbours, but provided that you avoid collecting them from places where they get sandy and full of pearls, which are unpleasantly crunchy, they are generally excellent. The same is true of the cockles that you dig from the sand (leave them in a bucket of seawater for a few hours; otherwise they are crunchy, too), and the magnificent razor shells, which give themselves up if you pour a strong salt solution down the keyhole-shaped burrows that give away their position at low tide. All the snail-like shellfish, from whelks to winkles, that you can find on the seashore are edible, if rather tough or tedious to eat. Shrimping is a traditional sport, and you can sometimes catch crabs at moorings by dropping a fish head or a leftover chop bone over the side; recover the line very slowly after it has lain on the bottom for a while and stand by with a net, because even a crab may feel that matters have gone far enough when it breaks surface.

With SCUBA you can do better still. Scallops, crabs, and even lobsters are relatively easy if you take a hook and a goody bag down with you, indeed, perhaps too easy, once you have a little practice, so that it verges on the unsporting. So don't be greedy. Take what you want to eat here and now, and on no account collect extras to give away or offer for sale. Other people have a living to make, and the animals themselves have no defense against this sort of predation.

The main point to realise is that practically everything in the sea is edible. Even the disgusting sea cucumbers (*cazze del mare* and other rude names) yield longitudinal muscles that are delicious; the eggs and sperm of sea urchins are fine in season; and it is even possible to relish sea squirts, the *violets* that the French devour in the probably mistaken belief that they are good for the libido.

From all of this you will recognise that my family and I are unashamedly omnivorous. Many people will hold that this is a reprehensible condition. We should be vegetarians and realise that animals have rights and that one of these is to be let well alone.

The two propositions are not necessarily linked. Convinced vegetarians may hold, simply, that it is better for you not to eat meat and/or—and this is a separate point—that it is selfish to do so, cropping the top end of a food chain when the world is full of people who would be only too glad to eat the grain that we uneconomically feed to our stock.

I rather doubt whether being a vegetarian is actually better for your health. Vegetarians, who generally think about their diet, tend to compare themselves with omnivores, most of whom don't. It is certainly better to be a vegetarian than to be exclusively carnivorous. A straight meat diet leaves the body short on carbohydrates and the gut low on fiber, and while all the evidence suggests that this is fine for hyenas, who deal with the fiber problem by chomping up the bones and the gristle, it is a hopeless diet for fastidious humans. So maybe I should opt for a vegetable diet. But then I reflect on the nature of my teeth. Herbivores don't have teeth like mine; they go for flat grinders that wear away and grow continuously throughout their lives. Nor do I have teeth appropriate to an outright carnivore, sharp things for slicing through flesh. What I have is a set of compromise teeth, not unlike those of a pig, not particularly good at cutting or grinding, but good, general-purpose gnashers for eating almost anything that comes to hand. My closest cousin, the chimpanzee (we share 98 percent of the same genes), has very similar teeth, and it eats anything from termites to monkeys, if it can catch them, to supplement an otherwise largely vegetarian diet. I conclude from my teeth (and its) that my digestive system has probably evolved to cope with an omnivorous diet and that a unilateral switch on my part is likely to result in the maltreatment of a delicate mechanism, tuned to run on a mixture of fuels that includes flesh as well as greenery. Most of us do, I suspect, eat a great deal too much meat when we get the chance, but that is another matter.

The selfishness of the human carnivore is a sound argument only for so long as it rests on the consumption of grain-fed stock or, stretching a point, of stock that eats grass on potentially arable land. It won't wash for sheep, or deer, which can extract a living from moorland that could never be arable. And it is nonsense when it comes to cropping the sea. There is no way of getting sensible amounts of food from the sea if you insist on a vegetarian diet. The tiny fringe of seaweeds round the edges of the oceans would make an almost negligible contribution, even if kelp were edible (actually we eat a lot of it—alginates, like chips, come with all junk foods—but they are tasteless and practically devoid of nourishment by themselves). The plants in the open sea are all unicellular and inseparable from the smaller planktonic animals, which they closely resemble as they swim about. It smacks of racism to eat people simply because they happen to be green.

The uncroppability of the plants in the sea means that vegetarianism in its more extreme forms is morally indefensible. Four-fifths of the earth's surface is sea. Why should anybody demand more than his fair share of the remaining fifth, or that fraction of the fifth that happens to be suitable for growing crops? The marine food chain culminating, inevitably, in bivalves and fish, crustaceans and cephalopods (squid and all that lot), is the only means that we have of realising the potential of the greater part of the globe's surface to generate food for people. By all means avoid terrestrial mammals (and let the sheep and the cows go for pet food, if that is what you want), but eat anything you can that comes out of the sea. It is only fair to the starving millions.

Animal rights is a far more complex matter. We seem to have no inhibitions about eating living organisms provided that they are green and firmly enclosed in little wooden boxes (plant cells all have cellulose cell walls). But we become confused as soon as they start to move about, and thoroughly mixed up if they even remotely resemble ourselves. As a biology teacher, I am very much aware that some of my students feel very strongly that it is wrong to kill animals, even worms and insects, for any, including instructional, purposes. Biology teaching does not involve much dissection these days, as it once did, but sometimes it is necessary to find out how creatures are constructed inside, and then we reach an impasse. The argument that the massacres of worms, snails, and insects carried out in the course of normal agricultural practice are on an altogether different scale from anything that we may perpetrate in a practical class cuts no ice with the principled young.

And not only the young. Many people are bitterly opposed to the use of any animal for any experiment or demonstration whatever. It is a point of view with which I have a great deal of sympathy—after all, I got into this trade because I find animals fascinating and aesthetically appealing. I am delighted to see that a regard for the well-being of animals is now built into the attitude of a great many sensible people who might never have thought about the matter a generation ago, as well as of the lunatic fringe who post letter bombs to surgeons practising on pigs, not people. But it would be hypocritical for somebody who probably owes his life to drugs evolved as a result of animal experiments in the past to vote for a total ban on this sort of thing in the future. I am as much as anybody in favour of alternatives to animal experiments, if such can be found, because some experiments undoubtedly do make animals suffer, and because they are expensive to carry out. Even drug companies have limited resources. I take issue with the more extreme antivivisectionists because they have fallen into the trap of believing that all problems allow for dichotomous solutions. You do, or you don't. The thing is right or wrong, and no cats are grey.

So let me try to explain the position of one practising vivisectionist (I do experiments on cephalopods). My thesis is that animals do indeed have rights, for a start with regard to other members of their own species. They also sometimes have rights with respect to some other species, and, arguably, we come into this when we domesticate animals. But there is no reasonable argument that can be put forward to support the notion that animals in general have rights with regard to other animals, and this includes ourselves. There are all sorts of good reasons for conserving animals and treating them nicely, but the argument for rights is the wrong grounds on which to tackle the matter.

An animal's rights begin with its relationship to the opposite sex of its own species. Any animal that has to copulate to reproduce has a right to assume predictable behavior in a potential mate, particularly if the said mate has teeth and a normally predatory disposition. The race would soon cease if this were not so. If the rules of courtship are simple, so much the better. If they are built in genetically, better still. Learning by trial and error is always a second-best answer if the environment is stable enough to select consistently for an appropriate instinctive behavior. Learning implies mistakes; misunderstandings can be fatal.

From recreation (as my mate puts it) to reproduction. Young animals are helpless. It is advantageous not to hurt them if they are likely to be related. It is best if the helplessness is obvious and the signal unambiguous. The puppy rolls on its back, exposing a vulnerable underbelly, whimpers or cringes in a decently submissive manner. The old dog is inhibited.

Both sides of this bargain have rights. The puppy has a right to assume inhibition. The old dog has a right to assume that any animal of the same species behaving like a puppy won't suddenly leap for the jugular.

This state of affairs can extend to other species in ways that have nothing to do with the survival of related individuals. Fish, to pick an example away from the higher vertebrates, have a contractual arrangement with certain invertebrates and other fish. Fish can't scratch themselves, so are liable to accumulate parasites, particularly around the inside of the mouth and on the gills, which they cannot rub off against the rocks or the weeds around them. They depend on cleaners, a range of shrimps and smaller fishes. The cleaner species are conspicuous in color and behavior, and often hang about in particular locations where the fish in need of grooming can rely on finding them. What the cleaners get out of it is a free meal, and the contract implies that the cleaned species is properly inhibited from making a meal of the cleaners, which are obliged to venture into some pretty dangerous places to do their jobs.

So do not let us argue whether animals have rights. Plainly, sometimes, they do. And in at least some cases, one can argue that one species has a right to expect predictable behavior from another.

One can extend this argument to cover our relationship with our domestic animals. In this case, deliberate selection of breeding stock has produced animals that behave in predictable and, to us, satisfactory ways. A cow doesn't run away when we want to milk it, and the family dog or cat doesn't try to eat the baby as soon as our back is turned. The quid pro quo is that these animals are to a greater or lesser extent dependent upon us. We've bred them to be that way, impressed upon them by selective breeding behavioral traits that would quite probably be detrimental or even fatal in the wild. We find ourselves in the cleanerfish situation, with some mutual obligations that we have to respect if the symbiosis is to continue. To be cruel to a dog that has had inhibitions bred into it so that it is reluctant to fight back is plainly in a different category from being cruel to a fox, which has no reason to expect anything other than unremitting hostility from our own species.

So I will back the proposition that our domestic animals have rights and even, perhaps, that imprisoned wild animals have rights, once they are inhibited by learning that humans feed them. But beyond this situation the question of rights becomes ridiculous. It is entirely the wrong grounds on which to defend the entirely right-minded proposition that we should treat animals with respect and even with love. There are plenty of good reasons for conserving animals and for treating them with what we conceive to be kindness that have nothing to do with supposed rights. Part of my job as a biology teacher is to try to convince people to recognize and respect these reasons, which range from aesthetics to the sensible management of resources. We should do these things from enlightened self-interest and respect for the future; no other reasons are necessary, and most other reasons make it extraordinarily difficult to be consistent. A blanket "Don't maltreat animals" won't allow me to swat a mosquito. I am reluctant to do that anyway. She (always she; male mosquitoes don't bite you) is a brave little thing, attacking against all the odds in an attempt to stash away a little protein, my blood, for the sake of her future children. But I kill her nevertheless because I don't like being bitten and I have no intention of carrying my views on animal conservation to the point at which I become a one-man nature reserve for the malarial parasite and a menace to my friends and relations. If there is a logical difference between my killing the mosquito, spraying the caterpillars that are busy devouring my lettuces, and shooting the rabbits ditto, I'd like to know what it is.

As a matter of fact, I, personally, would be reluctant to shoot a rabbit, because I doubt my ability to kill it cleanly and swiftly in this manner, and I dislike the idea of wounding anything. But I don't mind breaking a rabbit's neck if I find one with myxomatosis, or if one bolts into the net to avoid my son's ferret, because I know I can do this efficiently and quickly. So I have my inhibitions. I recognize these as personal quirks created by my imagination and my upbringing. Few people now live in daily contact with animals (I am constantly amazed at how few of my students have ever had to gut a chicken), and I wonder about the origins of the very strong feelings of many of the people who campaign most vigorously for animal rights.

Deliberate cruelty to animals of any sort disgusts all those whom I would regard as reasonable people, but the only logical argument that I can range against it, outside of the contractual obligations already discussed, is based on the argument that the sadist, having generated a taste for this sort of thing, may wish to move on to people. We should stop cruelty because it is degrading for the individual concerned and bad for society in general. We should try to educate people who are unwittingly cruel to animals, because it upsets our human sensibilities, not because the animals themselves necessarily have any right to expect better treatment.

It is worth reflecting that the rather odd idea that we should be nice to animals is by no means universal, even among educated people. The sweet little girl in my university class in Papua New Guinea, cheerfully nailing a struggling toad to a board, quoted the bible, "and have dominion over the fish of the sea and the birds of the air, and over every living thing that moveth upon the earth" (Genesis 1:28). When I remonstrated with her, she agreed reluctantly to let me destroy the toad's brain before she opened up its chest to examine the beating heart, a concession to my authority, not to my sanity. She was brought up in a society living a lot closer to nature than mine. Who am I to be so sure that my attitude is correct, or even sensible? Do I know more about seals or respect them more than do the Inuit? My own activity as I drive to work, spewing poisons into the atmosphere, is probably doing a good deal more harm to the creatures around me than is their seasonal whopping of a few seal pups over the head, a Spaniard's ritually slaughtering a bull, or a fisherman's bashing an octopus to a slow death on the rocks of a Mediterranean seashore.

Ponder these things, because they are important. But avoid principles. Principles are as often as not an excuse to stop thinking about matters that are difficult because they require judgment, and, as such, principles often constitute a sort of moral cowardice. We need animals, and animals, sometimes, need us. If we have to exploit animals because people will not otherwise pay to preserve them, so be it; it is the lesser of two evils. We must study animals, so that we can learn to live with them, and we must realise that if we allow a species to become extinct now, we shall never get it back. "Let it be" is a good starting point. But it is only a starting point, and it may result in extinction through inattention, whereas conserving the same species to eat it, or shoot it, or fish for it could ensure its continued existence. The important thing is to retain the huge range of creatures that we still have, by whatever means, until everybody has learned that animals are worth preserving simply because they are animals and beautiful and irreplaceable. Concentrate on that, and don't waste time campaigning against the seal cull or foxhunting or bullfights. These activities may seem repugnant, but they are not doing irreparable harm. Burning off the rainforest, draining the remaining wetlands, and poaching the last few rhinoceroses is. The world will be a very much duller place when these things are no longer here.

SUCCESS

People are confused about success. Most agree that they would like to be successful, but they all mean very different things by that. Television culture, to which one suspects most people consciously or unconsciously subscribe, equates success with the capacity to acquire possessions. Money. Up with the Joneses. Villa in the Caribbean. Second wife, young and glamorous, in middle age. Or influence, which might mean forgoing some of the above in order to retain the votes of people who are snobs about the television culture. The ability to do something very well, even if it didn't make money, might be admired, but if it was all that good, surely people would pay to see or hear it? Be a guru, bring comfort to your followers, and make a packet in the process would seem to be top job.

Biologists have an alternative definition, at least as regards success in other animals. It is called inclusive fitness. Inclusive fitness is quantified in terms of the number of viable offspring left to perpetuate the line—one's own offspring, or, failing that, offspring of close relatives. The important point is the passing of the blueprint, the genetic specification, from one generation to the next. Close relatives will carry many of the same genes, which is better than nothing. If potential immortality is not ensured in this way, the animal, or at least an individual's particular variety of the animal, becomes extinct, in which event whatever else it achieved during the course of its lifetime is pointless. What matters is the individual's blueprint, mine or yours, not the blueprint for the species. Natural selection doesn't give a hoot about the survival of species. Survival and multiplication are an individual matter; the representatives of the species that survive are the descendants of individuals.

Which is where bodies come into the matter. If a body that differs in some manner from those of the neighbors increases the chance of perpetuation of that particular blueprint, that sort of body will be more common in succeeding generations. This is what evolution is all about, the sole reason why life is no longer represented only by bacteria, or algae, or long-chain molecules multiplying in the primordial soup.

In general terms this means an inevitable increase in complexity and diversity, each slightly different body permitting exploitation of hitherto uncropped pastures. But because a body is also a sidetrack, a waste of resources that should properly (says natural selection) be dedicated to the propagation of yet more copies of the personal blueprint, there is a second tendency always at work: simplify, cut the overhead, drop off structures that don't contribute to the numbers of the next generation. A sessile body is better than a moving body if it yields a better chance of leaving offspring that survive to breed in their turn. Brains are extravagant. Plants are every bit as successful as animals. The race is not always to the swift, to the intelligent, to the well armored. Keep an eye on the balls (or the ovaries) is the one essential rule of nature's game.

So much for the short term. Look after tomorrow and immortality will take care of itself. Or will it? A feature of the way animals evolve is that natural selection is totally unaware of the future; the flavor of the month this month can lead to selection of bodies totally unsuited to next month's fashions. My tail is more splendid, attracts the females today (any idiot that can survive with a handicap like that must be tough); my splendid horns enable me to beat off the opposition, acquire a harem. Some of these advantages may be bought at a price that endangers the survival of the species (the tail makes me easier to catch; the horns may be fine for combating other elk, but not so hot in a fight with wolves, when shorter, spikier ones might be more effective), but that is irrelevant; what always matters is that I father more of the next generation than you do, even if the total number produced by the two of us declines as a result. A species can price itself out of existence as a result of internecine warfare.

Were the animals evolved en route therefore failures? The biologist's definition of success begins to fall apart as soon as he begins to consider the long haul. In the world around us we see thousands of animal species, all of which are successful in that they are here and now, surviving and reproducing. Humans and sea urchins, birds and bacteria, all up and running (or just sitting pat). And some have been running for a very long time indeed, an ancient aristocracy that, if sentient, might well be appalled by the elaborations that the higher animals have been obliged to adopt in order to perpetuate their kind. "Higher" is a bit of a laugh, at that—an attempt to redefine the way the world is so that we come out at the top of life's tree. A higher animal is more successful because we say it is. People, indeed, can live almost anywhere—like the insects, we don't do too well in the sea, and that is most of the surface of this planet, but let it pass—we can convert most other living matter, and some inorganic materials, into food, and we can screw up the ecology of a world if so minded, which we seem to be. But an intelligent sea urchin, let alone an intelligent bacterium, would point out that there are not all that many of us as a result and that our long-term prospects do not appear too rosy.

Colonize a few more planets and we might really be in business, a biological success at last. In the meantime, what price long-term survival of the genes? Some animals have survived apparently unchanged since the Palaeozoic—a couple of hundred million years or more, a good inning by anybody's standards. *Neopilina* (a limpetlike mollusc), cockroaches, and coelacanths, and a host of soft-bodied animals that have left no fossils, or only traces, tracks, and burrows, exist today in forms that are probably little altered since the first glorious explosion of diversity that followed the invention of multicellularity. Each is doing its own thing so well that it cannot be ousted, only outflanked by further elaboration to exploit ever more difficult alternative ways of making a living. Maybe the really successful animals are the ones that have survived without much modification, and that puts us nowhere.

A possibly more interesting proposition is that the biological definition no longer applies to *Homo sapiens,* because we can now achieve immortality in ways that have nothing to do with survival of the genes. I sit writing at a desk that belonged to my grandfather, and it was old when he owned it. The man who made it, a pupil, perhaps, of Thomas Chippendale—it isn't a certified piece of the master's work—may or may not have left descendants, but he produced a lovely artifact that will, barring accidents, still be around a thousand years from now. Maybe even more enduring are some of the ideas contained in the books that line the walls around me, or in the music ("West Side Story," as it happens), playing in the background. These things will maybe continue to affect human behavior and perhaps individual human inclusive fitness long after the genetic blueprints of their creators have been snuffed forever. It gives one ambition. It would be nice to create something immortal, and that arguably would be real success. It is one of the things that makes scientific research such an attractive career. "Politics is for the present. But an equation is for eternity" (Albert Einstein, in case you didn't know). By sorting something out, however trivial, one adds to the sum of human knowledge, and I like to believe that that accumulates. The contribution could be lost in some future dark age, but the odds are heavy that it will remain. Of course, it would be nice to feel that one's own contribution wasn't trivial, and scientists always believe that about their own little bits, but in reality you cannot tell what will prove seminal in the long run. Today's Cinderella of an idea—provided she gets to the ball at all—may emerge as the mother of a flock of little princes and princesses in a generation or two.

Meanwhile, I'm hedging my bets. My third grandchild was born last year. With two happily married sons, there is every reason to believe that the genes will go on. The criteria for inclusive fitness are satisfied, just about. And maybe one of these days I shall paint a picture or have an idea that the world will consider worth preserving.

MEASURING THE COST OF BEHAVIOR

To leave offspring, an animal must do many things besides find a mate. It has to find food and grow to maturity, avoid conditions that might snuff out or damage it, establish itself in the right place at the right time to ensure reproduction. There is only one criterion of success in animals, and that is measured in terms of the percentage contribution to the gene pool of the species.

Since all animal behavior is aimed at a single end, it should be possible to evaluate the cost-effectiveness of the different behaviors in which an animal is free to indulge. Given the choice, should the individual stay put and eat the grass here, or expend time and energy ambling round to the other side of the hill, where the grass may be greener? Should a rabbit dig half a dozen spare bolt-holes just in case a fox turns up between it and its usual burrow? Given the choice between a few large grubs and a lot of little ones, how should a foraging great tit behave? What proportion of its time should a lekking grouse spend displaying, vulnerable to predators but more likely to pull birds? *Do* animals always behave in ways that will optimize their reproductive success?

You need a common currency if you are to quantify the costs of choosing to behave in one way rather than another. One might, for example, use time. There are just so many hours available in a day or a lifetime, and time spent on any one activity means less time available for all the rest. But time alone is not very useful, because some activities plainly require a lot more energy than others. To obtain a useful figure for the cost of behaving in a particular way we need to know the cost per unit time *and* the total time that the animal spends carrying out that activity. Only then can we begin to consider the cost-effectiveness of different activities in relation to the single, reproductive, goal.

Energy comes from fuel. Food is fuel. One could try measuring the calorific value of food eaten (actually the easy thing is to look it up in a book—the calorific values of most common edible materials are pretty well known). Calorific intake is useful if we want to know energy expenditure over a period, provided we weigh the animal at the beginning and the end of the period that interests us to make sure that it hasn't gained or lost weight. With most vertebrates we are fairly safe in assuming that the animal's water content won't have changed much. We ought, too, to collect the animal's wastes, and deduct the remaining calorific value of these to get an accurate measure of the proportion of the food actually used. If we know about the costs of its other activities, weight gain compared with calorific intake can tell us the costs of growth. But because reserves may be stashed away in the body in good times to fuel activities later, it is unlikely to tell us the immediate cost of any short-term activity.

The common currency used to assess the costs of behavior in the short term is oxygen uptake. Most animals are aerobic; that is, they oxidize their fuels to squeeze out the maximum energy yield. Fortuitously, the fats, carbohydrates, and proteins all yield about the same amount of energy per milliliter of oxygen consumed to burn them. It comes out at close to twenty joules, or 4.8 calories per milliliter; a joule per second is a watt, and sometimes that is a convenient currency in which to express costs. A man on a rowing machine or an exercise bicycle can just about crank out enough power to light up a one-hundred-watt bulb; he will be using five milliliters of oxygen per second to do it.

Oxygen uptake is easy to measure; any zoology lab contains a variety of devices to monitor the oxygen content of the air or water around an animal. Know the volume of the aquarium or other container the animal is living in, or measure the through flow, and note the drop in oxygen content.

Interesting and sometimes counterintuititive facts arise from these studies. For a start, it turns out that by far the most expensive thing that warm-blooded animals do is keep themselves warm. Two-thirds of all the food we eat goes to central heating. If I stayed in bed instead of going to work, giving a lecture, cycling across to my college for lunch, pottering around labs and libraries—a fairly typical office worker's day—I could cut my running costs by only about 30 percent. If I play a game of squash, the overall effect is disappointingly small, because the furious activity lasts for only three-quarters of an hour. With such a high basal metabolic rate burning up fuel whether I run about or not, the effect is minimal, whatever it feels like afterward.

The next most expensive thing I can do is grow. Growth means feeding, breaking down food into small molecules that can be absorbed through the gut wall, reassembling the components to make more of me. Something on the whole to be avoided at my age, but vital to my granddaughter who is probably, right now, using some 20 percent of her not inconsiderable daily food intake, a direct cost to my daughter-in-law, to build her growing body.

My grandson, who is a bit older, spends most of his day playing. I don't have figures to tell me what it costs a small boy to create mayhem, but some years ago a member of the lab I work in did a study of the cost of play in kittens. Kittens, it turns out, spend two to three hours per day romping around, tumbling their brothers and sisters, pseudofighting with their mother. Rambunctious, energetic stuff. Conventional wisdom would have it that this is all essential training; costs incurred now will pay dividends later because the learned dexterity will make it easier to catch mice and leave the toms better equipped in the fight for females. So how big are the premiums? The sum that one has to do is straightforward. Oxygen uptake while playing minus oxygen uptake at rest times the length of time spent playing. Oxygen uptake measured by comparing the in- and outflows from the large box in which the cats are living; time from videotapes. It turns out that all that playing adds a paltry 4 percent to the daily energy budget. The essential training comes cheap; it would be worth cats' doing even if the subsequent advantages were marginal.

A much more serious 30 percent of the daily intake (of Catomeat and Whiskas in that particular study) disappeared into growth, adding a few grams to the growing kitten. And almost all the rest vanished in heat.

The lesson is that central heating so dominates the cost of living for mammals and birds that even apparently grossly energetic activities, such as running around and fighting, add comparatively little. Only in extreme cases, as when birds fly round the clock during migration, do these costs exceed the inescapable background costs of maintaining a high body temperature.

Not so with cold-blooded animals. Here the costs of behavior become dramatically more important. It costs about the same for a kilogram of lizard to run a hundred meters as it does for a kilogram of cat to do so—the answer is in the region of 60 milliliters of oxygen—but because the lizard has no central heating bills, the effects on its daily economy are hugely greater. The cat can keep moving around all day with little effect on its overall daily budget; the lizard must ration itself rather carefully in relation to the food available. And so must the overwhelming majority of animals, the frogs and fishes and all the moving invertebrates. For all these animals, the costs of chasing prey, policing a territory, or displaying to a mate can readily form a very high proportion of the cost of living. A male fighting fish will double its oxygen uptake with the effort of displaying at its own image in a mirror; in a real-life situation the reproductive rewards for all that energy expenditure must be considerable. It should come as no surprise that so many cold-blooded animals seem to spend an undue—by our standards—proportion of their lives apparently doing next to nothing. Conserving energy by cutting down on moving about is worth it for an ectotherm. A homeotherm can't afford to hang about; the fearsome demands of its central heating system mean that it will rapidly go downhill if it does so. The origin, perhaps, of the Puritan work ethic.

In cold-blooded animals the cost of growth replaces the cost of maintaining a high constant body temperature as the biggest single charge on the economy. Once again, the actual cost is not very different in warm- and cold-blooded animals. Basic biochemistry has not changed a great deal in the course of animal evolution, and in oxygen terms it costs about the same in food intake and in processing to grow a gram of cat as a gram of catfish. But the relative cost to the fish is much higher, because the background is so low.

Steady feeding and growth can double the oxygen uptake of an invertebrate. Indeed, it can raise it to several times the starting value if the feeding follows a period of starvation, because many cold-blooded animals cut back to very low tick-over rates when food is temporarily short. This is a better strategy than is expending a lot of energy seeking a scarce resource. Cold-blooded animals are opportunists, cashing in when the living is easy, sitting pat and doing nothing when conditions are less favorable. A bird has to migrate; a snail or a butterfly can simply hole up and wait for spring to come. Don't bother to feed the goldfish when you go on a week's holiday; but you must leave adequate supplies for the gerbils.

FOUL ORGANISMS

Marine fouling starts with the tendency for organic materials to stick to surfaces. Even the cleanest of oceans contains a hundred or two bacteria to the cubic millimeter, and the numbers in coastal waters can easily range into the tens of thousands. The actual number is irrelevant anyway, because bacteria with food to absorb—the organic materials—multiply with great rapidity, forming a film over the surface that consists partly of the bacteria themselves and partly of a sticky mass that they secrete. Pieces of organic and inorganic debris adhere to the slime, and a brisk little ecosystem builds up, with protozoa moving in to browse the bacteria. Green algal cells stick to and become embedded in the matrix, and here they photosynthesize, producing sugars in the sunlight, cashing in on the trapped fertilizers and adding fuel to the steady accumulation of unicellular animals.

So far so good; the outside of your boat is only a bit slimy. But it is only a matter of time before the larvae of larger organisms move in. Some, like barnacles, are plainly attracted by the smell of bacterial slime—it signals, one must suppose, that the area is habitable and ripe for development—while plants (the green strings of *Enteromorpha,* which form a fringe along the waterline) find that the film itself provides a soil for their tiny rootlets.

Antifouling paints delay, but do not prevent, this process. Most depend on the toxicity of heavy metals that diffuse slowly out of the paint. Antifouling is successful when the rate of diffusion of the poisons is only just sufficient to deter the colonists, so that it puts off the evil day for as long as possible. The trouble is, there are always some bacteria and some higher organisms—*Enteromorpha,* already mentioned is one such, and the small branching colonies of bryozoans (see later paragraphs for an inventory) are others—that are unusually resistant to heavy metals, and as soon as these hardy species gain a foothold the game is essentially lost. The mucilaginous gunk formed by the bacteria tends to cut down diffusion of poisons from the paint, and the bodies of the first colonists provide a substrate for the settlement of more fastidious species.

Early second-wave colonists include a range of small worms—the family name is Serpulidae—that form a tracery of chalky tubes, clinging tightly to the surface. The trick here is that while the swimming larvae are justifiably leery of antifouling paint, which will kill them, the settled worms are relatively insulated by their calcareous tubes. So it only requires a small flaw in the paint, or precolonisation by some tougher organism, to form a focus from which rosettes of tube worms can spread out over intact antifouling paint. Serpulids are filter feeders, extending a crown of ciliated tentacles when at work, and they like places where the water is stirred up so that a continuous supply of unfiltered, if sometimes rather thin, soup is wafted through the tentacles. Propellers, propeller shafts, and the skeg behind the propeller are favorite places, partly because the soup supply is better there, but also because the turbulence quickly pulls out the poisons from the paint.

Barnacles like the same places as serpulids, and for the same reasons. A further group of filter feeders, the sea squirts, or tunicates, is less attracted by the hurly-burly around the propeller and prefers dark overhangs. In our boat these particularly go for the slot where the drop-keel retracts. Tunicates come in several formats. The larger individuals, which can be centimeters long, are solitary, and resemble condoms or crumpled masses of wet cardboard. They eject a stream of water if you touch them; hence, "sea squirts." Smaller individuals are typically colonial, forming little star-shaped groups, often yellowish individuals in a blue- or reddish matrix.

And then there are hydroids and sponges and even sea anemones. And goose barnacles, and the larger brown seaweeds, besides a myriad of free-moving organisms now living in the shelter of the forests hanging from the hull. The rear-guard action has failed again, as it always does.

So there is a succession. Among the weeds, unicellular algae give way to filamentous and straplike forms that are eventually replaced by the brown seaweeds that cover our rocky shores. Antifouling paints slow the succession, but do not greatly alter it, so that within a year or two you find yourself in possession of a fine imitation of a moderately sheltered rocky coastline. Moderately sheltered, despite the fact that you sail it out into the open sea, because the boat always moves when a swell hits it, instead of remaining rigid and taking the full force of the waves. It is a nice thought that if one had the nerve to let well enough alone, the surface might eventually become colonized by limpets and sea urchins that would actually improve matters by browsing down the algal mat and preventing the settlement of further seedlings of the larger brown seaweeds, just as they do on the shore. The thought of a nicely balanced ecosystem, maintaining itself indefinitely, is appealing to a yachtsman zoologist—ecologically laudable and the green vote would no doubt approve. Be kind to sessile animals. They got there because they had no place else to go, and their needs are very simple. They are working hard to make the sea clean and sparkling. But they do have a deplorable effect on the speed of the boat. The skin friction of a carpet of barnacles is about four times that of a clean surface, so I have never quite had the courage to sit back and wait for the upturn.

In times past, fouling organisms had even more undesirable effects. Fiberglass is relatively indestructible, but wood isn't, and a number of animals liable to settle on hulls eat wood, with all sorts of lamentable consequences. Well-seasoned wood isn't all that nourishing, but will serve very nicely as a substrate for an animal with patience and bacteria in its gut. Besides, tunneling in wood provides shelter, so that once a wood-boring animal is established, it is made for life. Two sorts of wood-boring animal were largely responsible for the marine disasters that followed, a crustacean related to wood lice, *Limnoria lignorum,* the gribble, and a family of bivalve molluscs, of which *Teredo navalis,* the shipworm, is the best known.

It is debatable whether either of these organisms can actually digest wood. Cellulases, enzymes capable of breaking down chips of timber into absorbable sugars, are rather rare in animals, and most wood-eating animals depend on bacteria or protozoa with cellulases living in their guts. To prove the matter one way or another, one would have to rear shipworms or gribbles from their eggs under sterile conditions and induce them to infect carefully sterilized timber. To the best of my knowledge this has never been done. It matters little. In the real world, timber that has been soaked in seawater for any length of time is anything but sterile. It already sprouts a covering of bacteria and fungi, digesting the wood and waiting for any wood chewer to happen along and increase the surface for them to work on and/or provide a hospitable gut to house them while they digest.

Between them, the crustacean gribble and the molluscan shipworm and their symbionts can reduce a wooden boat to fragments within a few seasons, a plague the more dangerous because both organisms, for reasons of their own safety, prefer to work under cover, leaving a thin film of intact timber that conceals the evil inside until the whole structure abruptly crumbles. The only cure for these marine termites, before the days of poison paints, was to tar or copper or lead sheathe the hull (and unless the sheathing was perfect, the buggers got in through the cracks), move into freshwater from time to time, or haul out and dry off at intervals of a few weeks or months. And that opened up all the seams and made the boat leak.

Nobody is quite sure how the ancients coped with the problem. The pests were there, waiting, before man first provided them with the inestimable benefit of boats. They depended on driftwood, washed down the rivers from the then more abundant forests. The chances of finding a bit of driftwood were, one must suppose, quite small, but the animals made up for this by producing myriads of young, so that the sea was generally full of young hopefuls, paddling about on the off chance that they would hit upon a piece of timber before the reserves provided by their mothers ran out. If one made it once in a while the future of the species was assured.

The advent of wooden shipping and the tendency for ships to cluster together in ports must have heralded something of a golden age for marine termites. Naval vessels in classical times were generally hauled out at night—no great problem with a crew of a hundred or more rowers to a trireme—a necessary precaution to keep the hulls immaculate when speed and maneuverability determined who rammed whom. Commercial, load-carrying vessels were another matter. One can only assume that they were hauled out and scrubbed at intervals. The fact that sailing in the Mediterranean was generally restricted to the summer months—with no weather forecasting it was just too risky in the winter—probably meant that they spent long enough ashore or in freshwater each winter to kill off the borers.

It is interesting to reflect upon the extent to which history may have been determined by these subversive organisms. A blockading fleet was at an inherent disadvantage compared with the defenders, who had their triremes ashore and waiting. This is one reason the Punic Wars dragged on for so long; short of actually sealing off the other side's naval base, as the Romans eventually did at Carthage, there was no way of containing an enemy who only had to wait until you went home for a scrub or fell apart on station.

Throughout the Middle Ages it was, for practical purposes, impossible to keep a fleet at sea for any considerable length of time. Quite apart from the sailing qualities of the ships concerned—practically none could claw away from a lee shore until the evolution of the fully rigged ship in the late sixteenth century, and even then it was dodgy—a blockading squadron would soon be in no condition to chase a clean-hulled runner. Until the invention of reliable cannon, at about the same time, only the largest ships, with huge crews and mountainous topsides, could hope to defend themselves against an attack by galleys, and a dependency on galleys favored the side close to base. It is arguable that the defeat of the Spanish Armada, which took months to assemble, was as much due to fouling as to any inherent superiority in the speed and maneuverability of Drake and the lads. Only in Napoleonic times did it become possible to mount a successful blockade of a port at a distance, by relays of sailing ships with copper-bottomed hulls.

Nowadays the problem of antifouling is not so much the damage that the environment can do to your boat, as the harm that your boat can do to the ecosystem around it. Copper, in sheets or in paint, is not too bad. It leaches out into the water and kills animals if it is there in high concentrations, but in low quantities they can cope. Crustaceans and molluscs have copper-based blood pigments, just as we have iron-based hemoglobin, and their physiologies can pack away small excesses of copper into forms that effectively withdraw it from circulation. My octopuses eat a lot of copper with their prey and, so far as I can determine, never excrete any of it. Copper simply accumulates in the digestive gland, along with cadmium and some other heavy metals, in quantities that can rise to grams per kilogram. I avoid eating that bit.

Problems arise because we now have much more effective poisons to add to the paint, things that animals do not appear to be able to pack away and forget about. The one that has made the headlines in recent years is tributyl tin. TBT, an organic compound linked to a heavy metal, works a charm. My boat only needed antifouling every couple of years and was staying afloat all winter at that.

The trouble with TBT is that it is too good. It leaches out and poisons marine organisms that are minding their own business far away from the boat that is trying to repel boarders. It seems to be particularly bad for shellfish, and that is bad because oysters and suchlike are farmed in just the sort of sheltered estuaries that yachtsmen like to use for their moorings. The stuff is deadly at concentrations measured in nanograms per liter. A nanogram is 10^{-9} grams, one part in a thousand million.

Even at that dilution it has some pretty weird effects. *Nucella lapillus,* the dog whelk, was chosen the subject for investigation in England because it is, or was, found practically everywhere (it is the one- to three-centimeter-long white or color-banded snail that you find on seashore rocks, together with the barnacles or mussels that it eats for a living). *Nucella* is not immediately killed by low levels of TBT, but it stops reproducing. The females all begin to grow male genitalia, a development that blocks the egg production line so that the poor things end up eggbound and puzzled about their sex life. Ireland was the first country to ban the sale of TBT, in 1987. Within a year or two the ban was extended throughout the European Community. By then, dog whelks had practically disappeared from popular yachting havens. The ban was initially made only for boats under twenty-five meters long—pleasure yachts and the sort of coastal fishing boats likely to anchor in small estuaries. It didn't apply to aluminium hulls, on which copper-based antifouling was liable to produce electrolytic problems, and for a year or two there was a surprising proliferation of yachts pretending to be aluminium and hoping that nobody would look. Two years after the ban, the shelves of the chandler from whom I was then buying my antifouling in the south of France were stacked high with Italian-made paint that, on careful examination of the small print, turned out to be TBT based. Things have gotten better since then, but there is still (1998) no international law forbidding the use of TBT on ocean-going vessels. Supertankers still use it, and there are widespread reports of imposed sex changes from places as far apart as Sullom

Voe (an oil terminal in the Shetland Islands) and the Strait of Malacca. Significant amounts of TBT are now being reported in deep-sea sediments off Japan. There is a move towards an international ban within the next decade.

Now we wait to see what happens next. Nobody knows how quickly the toxic material will disappear from the system, trapped in sediments or simply diluted into the ocean at large. Preliminary information from areas where bans were imposed early on suggests we are talking in tens of years—yet another scientific breakthrough that seemed like a good idea at the time.

CRAWLING UP WALLS AND WALKING ON WATER

Archy the cockroach used to write material for the *New York Sun.* He did it, working nights, on Don Marquis's old mechanical typewriter, in the days before word processors. It is hard to work a typewriter if you are a cockroach. Archy, the soul of a poet reincarnated (one can only assume that he wasn't a very good poet in his previous life), had to do it by butting his head against the keys. He never managed to work the shift lock, because he wasn't heavy enough to raise the carriage. So capitals were out. The glorious emergence of CAPITALS AT LAST! was fortuitous; Mehitabel, the office cat, pounced on archy (hitherto always lower case), missed, but reset the shift. Mehitabel was heavy enough. There are some things you simply cannot do if you haven't got the mass.

It was bad enough for the soul of a poet to be entrapped in a cockroach, but consider the even more awesome possibility of reincarnation as an ant. Quite aside from possible resentment at the discipline (and sheer hard work) implicit in the social system of ants—all poets and cockroaches are liberals, like scientists, living off the leftovers of the comparatively rich—the poet would have problems with a physical world even more unfamiliar than the one Archy had to contend with. He, well, she, actually, given the genetic circumstances of most ants (you have to take the rough with the rough in reincarnation), would be obliged to live in a world without books, let alone typewriters, not because ants haven't the intellect to write books—we shall never know about that—but because books are physically impossible for them. The pages of a book small enough to be read by an ant would be inseparable, held together by molecular adhesion.

Nor could our artistic soul take refuge in painting or sculpture. Hammers and chisels don't work for ants; there isn't enough inertia in an ant-sized hammer. Painting would be difficult, too; watercolors out of the question, and there would be terrible troubles with the evaporation of more volatile solvents. A little unfired pottery, perhaps, but even here remembered techniques would be inappropriate. Water is a very different material at ant size. In small volumes it goes into lumps, rounds off to form droplets, held together by surface tension. An ant could never wash in the stuff, let alone slop it on with a paintbrush or use it to smooth clay, without first spitting in the stuff to reduce its surface tension. It could roll blobs of water about, maybe, though that is a trifle dangerous; if a foot breaks through it could be difficult to pull it out again. Just as well, really, that an ant, with its hydrofuge waxy carapace, cannot get wet on a rainy day, because it has no chance whatever of putting its six little feet up over a warm fire. Flames are unstable at ant sizes, and an ant couldn't get close enough to stuff fuel on a larger fire without frying itself.

There could be compensating advantages, more satisfying to the soul of a bodybuilder than to that of a poet. An ant can lift several times its own body weight, a potentially gratifying experience achieved because the force a muscle can exert depends on its cross section, not its volume. Volume is a cubic measurement, so the ratio of cross section to mass rises as you shrink. The enormous strength is an illusion; the ant's muscles are no stronger than yours or mine, it just happens to win out on the cross-section-to-volume ratio. Provided it doesn't topple over, it can lift objects larger than itself. An ant suddenly increased to human size would probably rupture itself attempting the familiar, but now impossible.

The same scaling determines other aspects of athletic performance. Walking or running uphill is much less exhausting for the small. The work that has to be done to raise a weight is exactly the same, gram for gram and centimeter for centimeter, whatever size you are. The mouse and I are subject to the same pull of gravity. The difference in the apparent ease with which we scramble up things arises because the mouse, being so tiny, ticks over much faster than I do; all its other costs—keeping warm, keeping its tiny heart beating, digesting its smaller but much more frequent meals—are greater gram for gram than mine (see "How Old Is a Fish?"). Against this background, the cost of raising a few grams of mouse a meter or two is relatively trivial. By the time you get down to the size of an ant, the whole question of vertical being somehow more laborious than horizontal becomes laughable (or would be; ants cannot laugh, a problem of sound wavelength in a tiny larynx, even if they had one); so up-down ceases to be different from sideways.

Falling is again no big deal for a small animal. I know a zoologist who dropped a mouse off a four-story building into a car park. He is a kind man, knows about scaling, and realised that the mouse (it was a smallish mouse) wasn't going to get hurt. In fact, the mouse did even better than he expected, apparently being well adapted to this sort of thing, though, one assumes, usually on a smaller scale. It stuck its legs out sideways, further increasing its area in the direction of descent, so that the terminal velocity it achieved was even slower than the scientist had calculated. It would have been cruel to toss a rabbit after it, and downright stupid to jump himself. A horse would have splashed, the ratio of mass compared with the surface taking the impact being altogether too great for flesh and bone. Jerry doesn't need to wear a crash helmet; Tom does. I wear one to ride a bicycle.

By the time you reach aphid sizes, falling is sometimes quite difficult. The least upcurrent of air can mean that an aphid, which may have thought that it was coming in to land, is wafted upwards faster than its rate of free fall. Join the aerial plankton instead; feed the swallows.

Safer to stay on the ground, sucking the sap out of my roses. Close to any surface, the air is hardly moving, whatever the conditions a foot or so up. Stick within a millimeter or so of the surface, and conditions are remarkably placid, even in a gale. The air immediately above the surface is stationary; successive layers slide over each other, with increasing velocity, but it may be a full meter up before you are subjected to anything approaching the speed of the wind as we feel it. Down in the boundary layer, close to the surface, it is warmer when the sun shines—no wind chill—and humidity changes little. The relatively even climate is a matter of no small importance in the evolution of animals—one reason for staying small, both on land and underwater. A fast-moving stream or a seashore with breaking waves is a dangerous place to be, but the chances of being swept away are much reduced if you are small enough to hunker down in the last few millimeters of the slow-moving boundary layer. Keep the head, tentacles, or whatever below the parapet.

And hang on. A very small animal needs only a very small crack or lump to cling to. The fly on your ceiling doesn't think the plaster is as smooth as you do, and it has the advantage already observed: the ratio of muscular force available to the weight to be carried makes hanging upside down no problem. Quite a few small animals can manage even if the surface really is smooth. They take advantage of the same phenomenon that would stop the ant reading a book; if very closely applied, the molecules forming the soles of the feet come close enough to those of the substrate for the two to attract one another. The contact becomes distinctly tacky, things need to be pulled apart to move. Even quite large animals, a few centimeters long, can play this trick. Next time you meet a gecko—the little nocturnal lizards that you meet in the tropics or around the Mediterranean are geckos; be nice to them because they eat mosquitoes and a lot of other insects—look carefully at its feet. It has flat little toes, which spread out over the surface, giving a lot of foot area. Underneath, each toe is covered in a sort of pile, like a very fine carpet; the skin of the sole of the foot is infinitely folded to form hundreds of thousands of tiny processes, each making very intimate contact with the surface below, sticking the animal even to something as smooth as glass. It is a fine trick if you are light enough that intermolecular forces can take the weight.

That goes for walking on water as well. In this case, the intermolecular forces concerned are those bonding fluids together. At the surface, the outermost molecules have nothing to hold onto except each other, so the net pull is inwards. Drops tend to round up, larger volumes flatten out under gravity but still minimize surface, so that if you poke at it with something unwettable, like a piece of wax, the surface dents in before breaking. Insects cash in on this; a thin wax layer is the means by which insects retain their body water, and, of course, it keeps the stuff out as well as in. Walking on water is an insect specialty, and many of them are light enough to be supported by the weight of water that they displace in the dents they make in the surface. Pond skaters are about as big as it is possible to be. They are careful not to take too many feet off the surface at any one time, they have efficiently hydrofuge toes, and they have changed their locomotor patterns to row across the surface. You have to be a specialist at this game, as other insects, crashing into the surface and getting stuck there because they are inadequately hydrofuge or unable to move away, soon find out. The only known open-water marine insect is a pond skater, a beast called *Halobates,* which can be found scooting across the surface of the oceans in the tropics.

Underwater the world of a very small organism can, again, be quite unlike that of its larger neighbors, for reasons that take us back to boundary layers. Animals come in all sizes, but physics does not. Water (like air) has viscosity; it sticks to most surfaces, and successive layers slide over one another. It takes force to cause the successive layers to do this, and this shows up as skin friction. All animals suffer from that; it is one of the factors that resist movement through air or water. The other is the inertia of the fluid that must be pushed aside to make way for the body of the animal. A big fish has a big volume, and pushing the water aside is the principal source of the drag that resists its movement. A small fish has the same problem pushing the fluid aside, but the situation has changed to a degree that depends on its size. The smaller you go, the greater the relative importance of skin friction—the familiar problem of surface-to-volume ratio, rising steadily as body size is reduced. There is an obvious effect of this on the shapes of animals. Big ones are streamlined, smoothing the flow of the fluid to be pushed aside. Little ones would be streamlined, but a smoothly tapered shape adds a lot of surface, and that increases skin friction. Small animals, down in millimeter sizes, appear to be distinctly casual about streamlining. At very small sizes, at fractions of a millimeter, the best shape to be is spherical.

Since water is water whatever the size of the animal struggling to move through it, the thickness of the boundary layer responsible for skin friction stays the same irrespective of scale. The only way to reduce it is to move faster, which thins the slow-moving layer at the cost of increasing shear forces in the successive layers of fluid. Skin friction rises. Small animals lose out because they cannot, in absolute terms, move very fast—if you look down a microscope and see them scooting around, recall that the microscope magnifies linear dimensions, but not time. So they are lumbered with a relatively enormous boundary layer—if anything, thicker than that surrounding larger organisms—stuck to them. It makes the world a very sticky place, like swimming in treacle. It gets worse as you get smaller. By the time you reach the dimensions of the smallest animals, measured in thousandths of a millimeter across, it is, as one zoologist (the same one who dropped the mouse off the building, as it happens) has put it, "like swimming in tar on a summer's day."

He is probably better at imagining such environments than I. The familiar becomes useless as a standard of comparison. Materials won't mix, because turbulence is impossible in treacle at any reasonable speed and temperature. Inertia ceases to be a force to reckon with. If you stop paddling, you stop, period. No coasting.

Once I experienced these things. But my multicellular memory carries no trace. The sperm that battled through to fertilize the ovum that became the other half of me went through it, once, just before my conception.

MACKEREL, *SCOMBER SCOMBRUS*: AN EXEMPLARY FISH

Birds are stupid because they fly so fast. Thinking is one of those activities that take a little time, easier for us slow, earthbound types than for anybody airborne who has to duck and weave to avoid abrupt endings. Elephants do not, I suspect, possess great intelligence, but, being ponderous, they have more time to reflect, something that cannot so readily happen to an intelligent mouse, living on a much faster time scale. Fish are not renowned for intelligence, and mackerel are fast fish. Life must be a succession of flash decisions as objects suddenly appear out of the murk: take it or leave it, so generally take it. It makes mackerel awfully easy to catch.

In summer, that is. People with yachts to the west of England start to catch mackerel in April and May as the sun brightens and the spring bloom works up through the food chain so that there are animals that mackerel find worth catching in the surface waters. Towing spinners or teams of bright feathers attached to hooks, we catch mainly the three-year-olds at thirty centimeters and more, fattening themselves after spawning in the spring and getting progressively plumper as the summer wears on.

Mackerel used to disappear in the winter. Traditionally, they were held to withdraw into deep water and bury themselves in the mud, like swallows. Both hypotheses have now fallen into disrepute. Conventional wisdom now holds that swallows fly off to Africa and mackerel retreat from the surface waters and hang about in vast shoals, miles wide and close to the bottom, offshore, in deep water, but definitely not buried in the mud, off the southwest of the British Isles. They don't feed much, live off the fats accumulated during the summer, and emerge again at the edge of the continental shelf in February and March, ripe and ready to spawn. Eggs and hatchlings are found in the plankton from then on, moving slowly eastward, and growing by the day on the plankton that blooms as day length increases. The tiny fish feed on yet tinier organisms and, surprisingly, to a considerable extent on the fecal pellets of the copepods, millimeter-long crustaceans that will themselves form the major food for the growing mackerel in the months to come. Growth is rapid, a millimeter or so per day, so that by the end of the first six months the young mackerel in the western approaches will measure ten or twenty centimeters long, and shoals of vigorous little fish set out to wreak havoc among the other inhabitants of the surface waters of the Channel, the Irish Sea, and north into the Hebrides.

Around the British Isles there appear, in fact, to be two distinct stocks of mackerel. One, as already mentioned, overwinters in the western approaches, spawns along the continental shelf, and then migrates annually up-channel and around the north of

Scotland to invade the North Sea during the summer. A second stock stays in the North Sea, spreading across the rich shallows before pulling back into the deep water off Norway when the southwestern stock is retreating to the region west of Cornwall. So far as breeding is concerned, they probably don't mix. The fish become sexually mature at a length of thirty centimeters or so, at the end of their third year. They may continue to spawn, migrate, and return to spawn again for a further ten or fifteen years. The annual rate of death from natural causes seems to be around 15 percent for fish that have survived the first dangerous two or three years.

Overwintering mackerel were relatively safe until the invention of sophisticated echo sounders. The postwar era saw the rapid deployment of fish-finding technology and, from the mid-sixties onwards, the very rapid development of a winter hand-lining fishery for the now readily located hibernating masses. Landings from the winter fishery soon overtook the catches from the roaming summer shoals.

By itself, this might have resulted in casualties that the remaining stock could survive. But the affairs of fish and men are linked, and the declaration of a fifty-mile fishing limit by the Icelandic government in 1972, extended to two hundred miles in 1977, spelled ruin thousands of miles away for the mackerel. Excluded from their traditional fishing grounds by the "cod war," trawlers from the British offshore fleet sought new places to fish. The first two boats moved in on the overwintering mackerel shoals in 1974, and by the end of 1977, the fishery had increased tenfold. U.K. landings alone topped 350,000 tons. Soviet trawlers joined the European fleets, and the future began to look distinctly bleak for mackerel. Pressure on the North Sea stock had practically wiped out the fish there by the end of the 1970s, and there was widespread panic that the larger, western supply of mackerel might go the same way.

Declaration of a two-hundred-mile fishing limit by the European Community in 1977 kept the Soviets out, but scarcely helped the mackerel, since Soviet factory ships remained and bought whatever was delivered alongside by the European fleets. Klondiking, it was called, a fisherman's gold rush. By the end of the decade, half a million tons of mackerel was being declared as landed (or sold to the Soviets) by European Community fishing fleets. God alone knows how many tons were sold but not declared. Most of the mackerel landed in the EEC went into pet food and fish meal.

It was at first believed that the Western European mackerel stock might stand this degree of predation. Surveys of the sizes and numbers of fish caught, recovery of fish tagged and released, and estimates of spawning success based on the number of eggs found in the spring plankton, led the mackerel working group of the International Council for the Exploration of the Sea (ICES, a scientific advisory body set up in 1902 in response to declining North Sea trawler catches—these problems are not new) to declare the breeding stock at around 3 million tons in 1977. An annual removal of 15 percent of the fish (added to the 15 percent natural mortality) should be sustainable.

But it wasn't. By the end of the decade, the stock had shrunk to 2.5 million tons, and by 1986 it was down to 1.5 million, a 50 percent decline within ten years. This, despite a variety of measures taken by national governments and the EEC, including a moratorium on trawling and purse seining off the Cornish and south Wales coasts (to protect young fish), agreement, in principle, to limit total EEC catches to the total allowable catch suggested by ICES (note the "in principle"), and a range of licensing schemes to share out the national quota among fishermen (greeted, one need hardly say, without enthusiasm). It seems inescapable that the actual numbers of fish killed must have greatly exceeded the landings declared.

This could happen for a number of reasons. One is simply that people cheat. It is just not possible to inspect and estimate every catch landed, let alone check on every batch of fish that has disappeared to countries outside the EEC. A second reason is that trawlers and purse seiners kill more fish than they land. The large, oceangoing boats process their catch, freezing it as it comes in; a sizable boat can handle between twenty and fifty tons a day. If it nets more than that, the excess must be shoveled back into the sea. It seems very unlikely that more than a tiny fraction of those returned survive. Mackerel have delicate skins, immediately damaged by any contact with a net. Released from the pocket of a purse seiner, they may swim away, but all the evidence from experiments with mackerel confined in nets, even under much less crowded conditions, indicates that they die soon after. They are, besides, ram-ventilated fish—the flow of water over the gills depends on the fish swimming rapidly forward with its mouth open. Jammed in the cod end of a trawl, it asphyxiates; dead or irreversibly stressed on arrival, it is pointless to put it back in the sea. One should note, too, that an excess catch may mean the entire catch. Big fish bring a higher price, and a fisherman with his hold nearly full or approaching the limit of his licensed quota will be tempted to dump the lot and try again if he thinks he may be able to catch larger members of the same species. There are some horrifying reports from SCUBA divers who have seen the bottom carpeted with dead fish, apparently in excess of requirements.

An even more fundamental problem in relation to the regulation of any fishery is that the parties concerned are never going to be able to agree on the object of the exercise. Even if one allows that all concerned wish the fishery to be sustainable (and that in itself may be questionable—why should a shipowner, with capital tied up here and now in a low-profit exercise, give a hoot about the yield a dozen years from now?), there is a problem about whether to shoot for maximum sustainable yield, most economic yield (most fish for least effort), or maximum employment (keep as many boats as possible at sea). However good for the public at large maximum sustainable yield might be, and however much individual boat owners might prefer most economic yield, both would require a considerable reduction in existing fleets—boats go downhill more slowly than do fish stocks—and that is politically unacceptable. Votes are lost, and the government has to pay the unemployed fishermen, the unemployed fish processors, and the unemployed from shipyards who build fishing boats. It is more economical in votes and taxes to keep on fishing until the public refuses to pay the high prices for the few remaining fish.

At the level of a single nation, it is important to be allocated as large a quota as possible. This is a further reason for the very rapid decline of the mackerel fisheries in the decade preceding 1984, when the EEC finally agreed on a common fisheries policy. Provided that it could do so without being caught in acts of total irresponsibility, every one of the states concerned was anxious to improve its starting position in the impending negotiations by establishing a total "traditional" catch at as high a level as possible. Blind eyes were doubtless turned. Eventually, ministers met and argued, and now do so annually. The collective pressure to raise the total recommended catch is immense; political expediency inevitably scores over scientific caution, and everybody prays for a succession of exceptionally successful spawnings that might pay off the overdraft. If all else fails, declining catches can be blamed on pollution; that is somebody else's department.

Any way you look at it, it is rough on the mackerel. But then, fish don't have votes. There has been some mending of international ways since the 1970s, when the fishery collapsed. Present evidence is that a crop of around 300,000 metric tons from the western stock might be sustainable. The problem is that fishing is now a highly scientific business. Every time we get to know more about the habits of mackerel and the distribution of the plankton upon which they feed, fishing fleets cash in. Legislation would have a hard time catching up even if there was the political will for it to do so. Sympathise next time you meet a mackerel, and at least do it the honor of cooking it in as delectable a way as possible.

GOD'S V.A.T.

I have never understood about physicists. Physicists hold that, other things being equal, random events will ensure that chaos eventually triumphs. Events inevitably run downhill into heat, and heat dissipates. The universe is doomed to depart with a scarcely audible whisper.

I do not understand physicists because they seem to believe that this is a long-term business, a matter almost of philosophy, or at least of years with a lot of naughts after, rather than a practical here-and-now affair. They have never worked in a marine biology station or had much to do with boats. They are not, you understand, in the forefront of the fight, battling against the sharp, cutting edge of the attack by the creeping plague (entropy gets into literature, too; unravel those metaphors). Entropy is now, it doesn't just concentrate in the rarefied atmospheres where physicists work. A broken pencil, a hiccough in a computer is their worst experience. They lead protected lives.

Marine biologists are familiar with unfamiliar life-forms. They probably would not recognise the arrival of an extraterrestrial, since it would hardly increase the range of the already present and peculiar. But they can spot a virus when they see one.

Consider carefully the following, all familiar manifestations in the laboratories that I visit to carry out my experiments. 1. Where does the electricity go? Why is the current unstable? Why does it rarely attain 240 volts, or even the 220 held to be delivered to the Continental labs that I parasitize? 2. Why does the plaster keep falling off the walls? 3. What is it that clogs up the pipes? 4. Why do everyday things, like clamps and clothes-pegs, wire, or ballpoint pens, degenerate into little heaps of corrosion as soon as one's back is turned? Seawater, you will say, seawater gets into the walls. Don't bleat about the obvious as though it were something abnormal.

But why not? My house, back in East Anglia on the edge of the fens, a soggy part of the world if ever there was one, isn't flooded. There are pipes, and the water, by and large, good-naturedly stays shut inside them. Electricity runs along wires, as it has long since been coerced into doing. In marine laboratories, something, let us face it, is subverting decent, well-behaved electrons and dihydrogen oxide to behave irresponsibly and leak out of carefully constructed wiring and plumbing.

As a biologist, I am, of course, familiar with the notion that viruses pervert DNA, and I am astonished to discover that my physicist colleagues seem to have no concept of similar self-seeking in the nonbiological world. Like tries to breed like. Influenza virus is in the business of converting little bits of you and me into influenza virus, and much havoc it wreaks in the process. It seems to me extraordinary that intelligent chaps like my physicist colleagues haven't tumbled to the fact that something analogous is happening to inorganic molecules. Worse, they are fatalistic about it, reluctant to tamper with what they perversely believe to be a law of nature. They invent new particles, as astronomers once did epicycles, in a desperate attempt to square theory with unwelcome facts. A defeatist attitude indeed.

In the presence of such ostrich-like behavior, what is a biological realist to do? To stamp out entropy is plainly a Herculean task, too much altogether to attempt as a one-man crusade. The stuff can remain dormant for years, like dry rot, and it is protected by vested interests that range from architects, who give special awards to each other for designing entropy reserves, as if entropy were an endangered species, through motorcar manufacturers, who are frankly dishonest about these matters, to the people who recently sold me double glazing, touchingly convinced that ultraviolet light won't ultimately convert their plastic into a breeding ground.

So what is the message for people as rightly concerned with self-preservation as yachtsmen? Very simple. Just face the facts and treat entropy like any other life-form. If the ends of your ropes unravel, set fire to them, and melt the ends. Flame sterilizes. If corrosion creeps up, spray it; WD-40 is a powerful entropicide. When anything electronic shows signs of erratic behavior, rap it smartly. Entropy won't be killed, but it may be cowed for a while while you search for dry joints, a favorite roosting place, with a soldering iron. The problem will likely be peripheral. Once it gets into the chips, cut and run.

All this, you will say, is what you do anyway. Of course. Entropy control is like the practice of medicine. You find out, empirically, what works, and then you seek a biologist to tell you why. All I am doing is providing a theoretical base, so that you may understand and systematize what you are doing. It is always a comfort to be logical.

The ultimate question is, naturally, whether we, collectively, have any hope of winning, even supposing that we can convert the traitors, the vested interests, who covertly support the spread of the disease. Will entropy ultimately and inevitably defeat us, as my gloomy physical colleagues would have me believe? I feel we should cry, "No!" for the dignity of the human race, if for no other reason. But there is hope. I rest my faith on history. Other living things have fought the plague for millions of years, and other living things have often won. We survivors are more complex than we used to be, still evolving in an environment dedicated to our extermination. The opposition is not invincible. Entropy lives and wins, most of the time. Most animal species are extinct, and it is worth reflecting on that, since that encourages caution. Entropy has had some spectacular successes in the past. It nearly beat us in the Permian, and at the end of the Cretaceous; it was a pity about the dinosaurs, after one hundred million years they nearly had it licked. But, by and large, increasingly sophisticated life-forms have outmaneuvered it. And that gives me faith, and is one good reason for being a biologist. One of us, at least aware of the problem, may yet stumble on a solution. After all, we beat small-pox.

I put all this to a colleague, a Canadian professor who meets me from time to time to play games with cephalopods. We were drinking beer in Papua New Guinea, in a break between attempts to fit up a refrigeration plant for an animal called *Nautilus,* which is one of the world's great survivors, an ally that has held entropy at bay since the Palaeozoic (but that is another story). He—the professor, not the nautilus (we have not learned to communicate with them yet)—said that all I have written above is nonsense. "Entropy," he said, plonking his empty beer can on the bench, "is God's value-added tax." There is, he averred, a tax on change of whatever sort, extracted and removed from the system. The idea is appealing. It has grandeur and elegance, a universal hypothesis that fits the facts. It would be neat if true, and that sort of simplicity always appeals to a scientist.

Except when one stops to think about it. The consequences of such a possibility are disturbing. What, you may ask, is God proposing to do with the accumulated revenue? Why is he (she) clawing back a percentage from his (her) own creation? I submit that the only possible conclusion is that God has doubts, wants to hedge his bets, accumulate capital against the day when he has to revamp a creation that has somehow failed to live up to his expectations. And as a biologist, I cannot help but reflect upon the past. From time to time the Almighty *has* rejigged the creatures of this planet (and God only knows what he is up to elsewhere), with catastrophic results to those—the ammonites, the dinosaurs, and almost everybody who lived in the Permian—who probably believed that they were doing all right in a world that they understood and regarded as predictable. Beware the next time; it could be you that is for the high jump.

Another good reason to fight down entropy. If you reduce the flow, God almighty may not be tempted to cash in his profits and start again. At least, not in *my* lifetime.

ABOUT OCTOPUSES

Once upon a time I was an entomologist. When the RAF let me go after an undistinguished career as a radar operator (I did no irreparable harm to Central Germany G chain and the Berlin airlift), I went to university to cash in a scholarship that I had won during my school days. His Majesty, who didn't seem to bear any ill will about my largely idle RAF career, was prepared to pay the fees. In those days, apart from young women, the thing that mattered most in my life was fly-fishing, and I cherished the faint hope that I could somehow combine this with research in zoology. Things didn't work out that way, and a year after I had begun a postgraduate course that might have led to a Ph.D. in insect physiology, I was offered a job on the staff of the zoological station at Naples, an eccentric enterprise in those days, run by a director who had inherited the business from his Victorian father (and in due course passed it on to his son—I worked for both of them). They wanted someone whose native language was English to internationalize their staff, help with foreign visitors, sort out the chaos still remaining after the war, and correct the English in the lab's journal.

It seemed an interesting thing to take on; I'd never been to Italy. My professor said it was academic suicide.

Joyce and I got married on the strength of it, chucked in two potential Ph.D.'s (she might have been an entomologist, too), and took the train, phrase books in hand.

The relevant part of this story is that there didn't seem much point in being entomologists at a marine biology station. So in the periods when I wasn't confusing the laboratory's fishermen with my awful Italian or soothing visitors who couldn't understand the language either, I searched around for something that might save me from academic suicide. If he emigrates to foreign parts and doesn't want it to be terminal, a scientist must publish the sort of papers that will be noticed by potential employers and maintain a never-to-be-forgotten campaign at conferences whenever he goes home on leave.

By far the most attractive animals in Naples were octopuses. If the only octopus that you have ever seen was a pale dead thing on a fishmonger's slab, you may be understandably deluded about this matter. The live animal is something else. An octopus in good form has a look of alert intelligence not shared by other marine animals. It changes color, skin patterns, and skin texture continuously, the more so if it is interested in what is going on around it. In aquaria, they rapidly lose their nervousness of people and learn to come and be fed. Toss a crab into the tank, and the octopus bobs its head, ranging the prey by parallax, flushes darkly (camouflage marred by passion), and makes a jet-propelled pounce. Smothering the unfortunate crab in the web of skin between its arms, it carries the food home. An octopus likes to have a hole to sit in, and it will gather stones and other debris to form a sort of nest if nothing more comfortable offers. We gave ours flowerpots to sit in.

Obviously, we had to work on octopuses. Note the "we." My wife is a feminist, but she is also a realist, and knows that two fairly clever people pulling together stand a better-than-even chance of beating any one clever person working on his or her own. It is one of the reasons marriage is a popular institution throughout the animal world. Another essay, that we will sidestep for now.

By great good luck, we hit on the octopus at a time in the mid-'50s when people were searching for animals less complex than mammals, when attempts to establish the relation between learning and the changes in nerve structure believed to be associated with learning had run into something of a morass. Octopuses plainly learned. They had brains much smaller than those of mammals, and they could hardly be more distantly related. This was important. Individual nerve cells, so far as we know, work in much the same way in all animals. To make a learning machine out of nerve cells implies fitting them together in a pattern such that their individual properties will collectively form a system that changes with experience. Just searching in any one sort of brain for a pattern that might do this is a bit like sorting through a haystack for a needle, with no clear idea of what a needle ought to look like. A brain does so many things, and learning is just one of them. But if the same patterns turn up in parts of unrelated brains each known to be necessary for learning. . . there was at least a hope. Besides, the American air force wanted to know how to build tiny computers capable of recognising patterns and homing in on them (the future cruise missile), and that was just what octopuses were good at. So the American taxpayer paid the bills.

Nice try. Several of us, led by J. Z. Young, from London (a man who ought to have gotten a Nobel Prize, but never did—he discovered giant nerve fibers in squid, a discovery at the base of most of what we know about how nerves work), learned a lot of things about learning in octopuses. Such as, they learn simple visual discriminations as quickly as do cats or dogs; they are colorblind, but able to recognise the plane of polarization of light (which we cannot); and they live in a geometrically odd world in which a sphere is a flat surface and a rod is a cube. Fascinating stuff. Touch learning can be chased into a tiny part of the octopus brain, less than one cubic millimeter, where memories are stored. And there we stuck. A one-millimeter cube still contains several million nerve cells, enough to find any pattern you care to look for, too many patterns to analyze with the techniques then available. So we went our several ways, and almost everybody forgot about touch learning in the octopus. In research, as in everything else, fashions change.

But learning in the octopus had served me well. I came back to a research fellowship at Trinity College, Cambridge, and a year later picked up a demonstratorship, first rung in the ladder, at my old department. I was thirty-one. My professor (the academic suicide one) congratulated himself; we send these bright young men abroad, he said, to prove themselves. Sir James Grays, the Epstein bust in our library, looks just like him. Academic suicide was temporarily averted. Like a lot of academics, I didn't get a permanent position until I was nearing forty. The theory is that if you survive to that age in this business you must be a workaholic and can be left to develop your own neurosis without the help of a sword hanging over the balding scalp.

Long before that I was hooked. Cephalopods—the octopuses, squid, and cuttlefish, and beasts such as *Sepiola,* after which our boat is named—are a grossly neglected group of animals, and I have been doing my best over the years to remedy this, working on a variety of topics ranging from the hormonal control of sexual maturity (they have a system not unlike our own) to cardiovascular performance (octopuses don't have heart attacks). The widespread neglect of cephalopods by the scientific community is largely a result of Anglo-American bias. Wasps (white Anglo-Saxon Protestants, in case you have forgotten) don't eat cephalopods, and their fisheries are largely concerned with the more favored fish (whoever talked of a squiddery?). We know a lot about herrings and plaice, but who is going to throw good dollars away on a study of creatures that only foreigners are crazy enough to eat? Many parts of the world know better. Cephalopods are not so underrated in the East or in Mediterranean countries. And this is what we shall all be squabbling about if the present overexploitation of fish stocks continues.

Evidence is steadily accumulating for the existence of massive numbers of oceanic squid and a huge tonnage of octopuses and cuttlefish in the shallower waters of the continental shelves. In the past, we have not been aware of these large numbers, because techniques that catch fish do not necessarily catch cephalopods. It is difficult to hook a large, soft-bodied animal; the hook tears out. It is difficult to net an animal that sees as well as a cephalopod, the more so precisely because the animal's viewpoint is not dominated by its equivalent of the fish's lateral line. The lateral line of a fish detects obstructions at a distance, stops fish from careening into things. But it also allows us to fool them with nets that they could perfectly well swim through if they believed their eyes rather than their pressure sensors. Only the cod end of a trawl is fine enough to prevent escape by most of the fish that find themselves in it. Most of them have been conned by the wide-mesh wings of the trawl. Squid, unlike fish, believe the evidence of their excellent eyes, rather than their pressure detectors, and swim through the mesh while the going is good. To catch such wary creatures you need highly specialised equipment, with enormous seines, or jigging ships that offer huge numbers of lures for the squid to attack—vehicles that cannot be switched to alternative fisheries if the squid crop fails, which, as we shall see, it is liable to do if not rather carefully managed.

Factory ships, collecting and processing squid, now roam the South Polar seas (arguments about fishing quotas around the Falklands are mostly about squid, not fish) and the North Pacific. They are all hunting shallow or midwater species. Deep-water squid remain as yet untouched. The surviving sperm whales alone are estimated to take a greater tonnage of deep-water squid per annum than the entire worldwide human catch of fish. Their prey includes species that we cannot catch at all. *Architeuthis,* from the northeast Atlantic, has a body as big as a cow (or at least a good-sized pig), but no man has ever caught one that was not already moribund. We know the toothed whales take them, because the beaks turn up in their stomachs, but our own sampling is limited to the collection of a few alarmingly rotten specimens from the beaches of Newfoundland and, once in a long while, from Norway.

So, there is lots of good protein, most of it immediately edible (there are some difficulties with the sort of squid that sperm whales eat, which are full of ammonia, but there is probably nothing that a modern processing plant couldn't render into fish—check, squidburgers—if we could figure how to catch them), croppable, it now seems, at rates well in excess of those applicable to most fish stocks. In the last decade or so we have come to realise that cephalopods are uniquely short-lived, as big animals go.

All the cephalopods that we fish for commercially grow like smoke, often from a few milligrams to a kilogram or more within a year; they mate, lay eggs, and die. The immediate cause of death was one of the first fruits of our hormonal study; once switched on, the hormone driving the animals into sexual maturity cannot be switched off. All resources are channeled into the production of eggs and sperm; the bodies waste and die. Turnover is staggeringly quick compared with fish, which typically take several years to reach maturity. So the potential crop is massive. The bad news is that it is also very variable. A single unfavorable spawning or feeding season can decimate the stock over a wide area. This is bad for profits and leads to overfishing at just the moment when the survivors would be best left alone.

All that, then, is background, the reasons I give when asked why I work on an apparently exotic group instead of on something obviously useful, such as mammalian hormones, cancer, or insecticides. I like to think that I am ahead of the game, sorting out the physiology of animals that everybody is going to have to know about one of these days.

Not, apparently, quite yet. Nobody has yet asked permission to nail a plate (Martin Wells lived here) to my door, and the queen is being coy about the knighthood. Not that it matters, because in the meantime there is enough interest to pay to get me around the world, and these animals do, wisely, and fortunately for myself, live in some very attractive places. The mate comes too, when she can; her career has led away down an alternative path, student counseling and administering at a Cambridge college, so that my only chance of getting her into a lab is to snatch her away to foreign parts, out of reach of telephones. Neither biological research nor college administration pays very handsomely, but when you come down to it, it is a very marvelous thing that the world is prepared to pay at all for the likes of us to spend so much of our lives in the study of anything so inherently interesting and outright beautiful as animals. In the bad times, I try to reflect on that. And there are bad times, just as in any other creative activity. Research is like painting pictures. The product hardly ever turns out quite as well as one might have hoped; it can be maddeningly frustrating; and one spends a lot of time simply cleaning up the equipment. But once in a long while everything goes really well, and that is euphoric. And even in the bad times one is adding something, however slight, to the sum of human knowledge. Some poor people work just as hard and all they make is money.

So that is what I do for a living, that and teaching. The research has gone well enough over the years that I now have an *ad hominem* appointment, with no duties other than to do whatever research I fancy; I teach because I like teaching, not because I have to. Whoever invented these jobs was shrewd; I probably work far harder because I am responsible to myself than I ever would under anybody else's direction.

Enough of generalities and self-justification. What should an honest amateur know about the octopus?

First, where to look. With mask and snorkel, you can find them in as little as a meter or so of water. Octopuses like to sit in holes, looking out, so that often you will see only the head and eyes, with the arms and suckers tucked away out of sight beneath the animal. Sometimes you can locate them from the piles of debris, shellfish and crab carapaces, forming a little midden outside their homes. To begin with, you will pass most of them, unaware that you are being watched; it is a matter of getting an eye in for the eye. Sometimes, as you pass, the beast will blow a puff of ink, and if you see a puff of ink, search the area carefully. As you come close, the animal may suddenly reveal itself by turning the suckers towards you, raising the arms to defend itself.

Or swim with a torch at night, shining it on the rocks and sandy patches between. Octopuses are most likely to be roaming about in the gloaming and are easy to spot then because they react to the spotlight by trying to scare you off, as they would a marauding fish. By flattening out and extending the web between the arms, and paling all over, an octopus can contrive to look enormously larger than it really is. The effect is enhanced by dark rings around the already prominent eyes. I would defy anybody not to back off when first confronted by this apparition; it is truly an alarming sight, the more so because everything appears larger when seen through a face mask. The pause is, of course, exactly what the octopus needs. While you recover, it changes color and shape and makes a jet-propelled escape behind a smoke screen of ink. The trick must work for fish, too, since it is hardly credible that such a defense could have evolved since men took to face masks.

If you see one of these amazing creatures and have an urge to bring it aboard for a closer acquaintance, be wary of the fact than an octopus has a lot of grip. Don't try to land a big one. Anything much more than a hand span across is potentially dangerous unless you know exactly what you are doing, not because it intends you any harm, but because its immediate reaction when grasped is to hang onto everything within reach and clamp down. If the objects grasped include you and the seafloor, you have problems; quite a small octopus could hold a snorkeler underwater.

If you do find a small one, the odds are that it will be in a hole. Don't try to get it out by poking. It will stay there, scared out of its wits, until you have damaged it so badly that you will, or should, hate yourself for what you have done. The answer is a squeeze bottle, the sort that washing-up liquids come in. Make up a very strong salt solution (dilute copper sulfate is better, but I can think of no good reason why any yacht other than ours should carry the stuff). Squirt that into the hole and stand back. The octopus will huff and puff and then, as often as not, make a run for it. Wait until he (or she, they come in the usual two sorts) is well clear of the hole and dive down, grasping the animal by putting your hand over the head, with the fingers between the arms. If you are reasonably deft and catch the animal by surprise, you will have it off the bottom and into the bag, which you forgot to take with you, before it figures out what is going on and settles down to any serious resistance. Otherwise, you will rapidly discover that it is difficult, nay, in the short term, almost impossible, to disentangle an octopus from your hands. The animal will cloud the whole issue with abundant ink, and the bag and squeeze bottle will drift away. Up to your neck in alligators when all you set out to do was drain the swamp. Your swimming companions will be working out witty histories of the event to embarrass you at some future social event.

If and when you get back to the boat with your new friend, put it in a bucket. He or she will climb out, proving a simple mathematical fact that eight arms can loop over the sides of a bucket quicker than two arms can put them back. The resulting struggle will lead to resentment on both sides, the more so because the octopus, if it has any left, will now squirt ink all over your clothes and the clean teakwood in the cockpit. Octopus ink is very difficult to remove. If you finally succeed in getting a lid onto the bucket, don't forget to take it off and replenish the water every ten minutes or so, or it will die, and then, in conscience, you are obliged to eat it.

Another word of warning. It may bite you. The little ones seem to try most often, and the nip that they give with the beak—

in the middle of the rosette of arms—rarely breaks the skin. The frightening thing is that you cannot get the animal off in a hurry. The worst reaction I have seen was that of a student whose arm swelled up as if she had been stung by a wasp; it went down in a few hours. But you could be unlucky; sometimes people are hospitalized by bee stings, and the saliva of an octopus contains a witches' brew of digestive enzymes and substances designed to knock out crabs that I, for one, would rather not have in my bloodstream. Note, too, that my own experience—I have been bitten a dozen or so times in thirty years—is limited to the common *Octopus vulgaris.* I wouldn't vouch for any other species, and at least one, the blue-ringed octopus (which, happily you don't find in Europe; it lives off Australia), produces a nerve poison called tetrodotoxin that really does kill people once in a while.

So the message is, let it be. If you are lucky enough to see an octopus while swimming, mark down where it is hiding and go catch a crab. Approach carefully along the surface and drop the crab close to the octopus. If you have been reasonably circumspect in your approach, the odds are that the octopus will dash out and collar the crab, and that is a sight worth seeing. It is always nice to do something that is so obviously appreciated.

LIFE WITH A LIVING FOSSIL

Christmas 1983; Dick blew into my lab, dressed like an advertisement for port, sporting a cloak and broad-brimmed hat. I had not seen him for several years, since he quit Cambridge, bored with ivory towers and full of good socialist principles, intending at that time to emigrate to Cuba in a boat he was reconstructing in his backyard.

The Cubans wouldn't have him; he was the wrong sort of left wing, I gather. And the boat was so palpably unseaworthy that even Dick, who is brave but not suicidal, thought better of it. And vanished.

Until now. "Martin," he said, "would you like a trip to New Guinea?" The resurfaced Dr. Moreton was now professor of biology, University of Papua New Guinea. A pillar of the establishment, as it turned out, a new facet of the man I recall as a hippy, wearing parti-colored trousers and carrying a handbag.

One of the guiding principles to which I subscribe is never turn down an invitation to go places with interesting animals. This dictum has paid off; Australia and Canada, Carolina and Hawaii, India, Ghana and East Africa, nearly all temporary teaching jobs. If you live and work in a place, even for a couple of months, you see a lot more of the detail than you can pick up on a flying visit as a tourist or attending a conference. And now, miraculously, Papua New Guinea was on offer.

New Guinea is something special for zoologists, one of the great watersheds, a country where the fauna suddenly changes. In Australasia, on the marsupial side of Wallace's line (the Wallace who might have written the *Origin of Species* if he hadn't been too busy earning his living and a little bit potty to boot), it is a country with near impassable mountains, birds-of-paradise, bird-wing butterflies, and seven hundred languages, and students whose grandparents and sometimes parents used stone axes and occasionally ate their neighbours. "We have," he added, "a nice little marine biology station a few miles down the coast; I could fix your teaching to give you long weekends at that."

Having agreed forthwith to go for three months, I asked what the deal was. One of his staff was going on sabbatical, and my brief was to create a comparative physiology course for the final-year class, help demonstrate practicals in earlier years, and carry out such small-group teaching as fell to hand. I would have agreed to almost anything. I even mortgaged my wife. She came out to join me for her Easter vacation. On the way from the airport I broke it to her that she had a class next day, a seminar and discussion group for twenty or so freshmen students, a mixture of highlanders and Papuans, on the physiology and general inadvisability of getting very drunk—something of a problem in Papua New Guinea because, amazingly, the mainland had no alcoholic drink before white man arrived. There was a handout on cirrhosis and brewer's droop; beyond that she was to play it by ear. She took it rather well, I thought.

She was a little more unnerved by the hostel where we stayed. Our room had a steel door and heavy metal mesh over the windows. A notice stated that one should on no account go out at night; if attacked, blow the whistle. The peg was there, but somebody had nicked the whistle. It is not clear who would have come; perhaps the caretaker would have rung the police, who drive about in armored cars. The country's capital, Port Moresby, is a dump, and at night rather a dangerous dump at that. Even on campus, houses have high wire fences, dogs, steel doors, and radio intercoms. The hostel for female students is surrounded by three banks of barbed wire, to keep out the roving gangs of rascals who, finding that the streets of the capital are not after all paved with gold, look for compensations. Spears are discouraged in town, but citizens carry pangas and other agricultural instruments which are almost as lethal. The campus is full of trees and flowers, birds in the courtyards, and tree frogs croaking in the walkways.

A first trip to the island; truck along the coast road, dirt track, and a dock where a boat is left for us. Poles marking the reef edges, roosting places for terns that stare as we circumnavigate. A turtle departs in panic; the coral glows beneath the surface.

The tide is out, and the boat grounds on the sandbar. Throw out the anchor, wade out, and collect the boat later. Step out carefully; the stingrays come into the shallows mostly at night, but there could always be the odd drowsy one. Carry the diving gear, food for the weekend, and things for the lab over the blistering sand.

Motupore Island Research Department boasts a laboratory, open to the four winds (which is good), but also to the rain, which blows in horizontally during the not-infrequent squalls (so all the more delicate apparatus is placed in upturned glass fish tanks). There is running seawater and electricity during the day from a diesel generator, and a rackety compressor for charging dive bottles. The tools hung in the workshop include shark hooks of terrifying dimensions.

Accommodation is in a bunkhouse on stilts, with a kitchen, and showers, provided there is enough reserve in the rainwater tanks. It has efficient mosquito screening, a matter of supreme importance, as one begins to realise after working in the lab, which hasn't. There the beasts swarm below the benches, attacking in tight formation, scarcely discouraged by insect repellent which drips off anyway with the sweat, because this place, even given perpetual sea breezes, is hot. After a few days, a newcomer looks like a Dalmatian. You swallow malaria pills and hope for the best.

The reefs are magnificent. Swim out from the lab—at high water the sea is ten feet from the door—into a snowstorm of fish over the coral. Only, each flake is a different color. The sheer variety is mind-blowing. What do they all do for a living? Below the fish lies a carpet of invertebrates from the coral heads down onto the sand. Don't touch; some of these, and a few of the fish, sting; a competent zoologist ought to know which.

Three miles out to sea is the barrier, the ultimate reef. It runs along the south coast of New Guinea, arcs round the Torres Strait, and carries on for a thousand miles down the coast of Australia, the most enormous structure that animals, ourselves included, have ever made. A wall of coral, vertical on the seaward side for the first hundred feet or so, tailing off into scree that runs out of sight into the blue depths. If the fringing reefs are mind-blowing, this one is a heart stopper.

In the depths beyond lives *Nautilus.*

The pearly *Nautilus* (in biology there is a convention: when a Latin name is used it is printed in italics, a common name is not italicized; "Nautilus" is used both ways) is an almost mythological animal, an invertebrate unicorn. It has the most beautiful shell in the world, a geometrically perfect spiral with a surface like bone china, decorated with flames. Aristotle knew about it, and stated that it had extendable arms, a tantalizing statement, because it seems in the highest degree improbable that he ever saw a living specimen. The nearest live location is in the Andaman Islands, east of India. Shells must have been traded westwards then, and continued to turn up in Europe throughout the Middle Ages, where they were regarded as great treasures and often made into elaborate ornaments. With the opening of the eastern trade routes in the mid-1500s, specimens became more frequently available, so that even a naturalist might aspire to possess one. The name seems to have originated about then. One hundred years later Robert Hooke reported on the animal to the fellows of the Royal Society and suggested that the chambered shell was a buoyancy device as well as a retreat for the animal, which he had never seen. The first description of the soft parts came in 1705, from one Georgius Rumphius, a Dutchman employed by the East India Company. His pictures were not particularly accurate, but sufficient to excite anatomists who were even then beginning to exploit comparison as a way of establishing relatedness in animals. *Nautilus* was plainly related to squid and octopuses, but it had an external shell and far too many arms; it was already recognised that the shell resembled that of many fossil forms, which were thus recognized as cephalopods rather than snails.

It was more than one hundred years before the first almost complete specimen, preserved in spirits (probably rum), came to

Europe and was delivered to the famous Richard Owen, premier anatomist of his time (the man *inter alia* responsible for reconstructing the iguanodon and other dinosaurs for the great Exhibition of 1851). He wrote a monograph in 1834. Knowing about the anatomy was not, however, sufficient to satisfy post-Darwinian biologists. By the end of the century, it was recognised that clues to the evolutionary past of an animal could often be found in its embryology: "Ontogeny recapitulates phylogeny." "Every animal in the course of its development climbs up its own family tree." What they all wanted was developing eggs.

Rather a tall order, for an animal found only in deep water in the then still quite remote South Pacific. Arthur Willey, a pupil of Ray Lankester, director of the British Natural History Museum, was awarded a fellowship at Cambridge and dispatched to collect some. His voyage lasted from 1894 to 1897, an epic, under conditions that makes one wonder how he managed to survive, let alone carry out serious scientific research. He, too, published a monograph, in 1902. He never did find any fertile eggs and, indeed, these were unknown until some fifteen years ago when, at last, *Nautilus* bred in the aquarium at Hawaii.

Following Willey, the biology of *Nautilus* was essentially neglected until the 1960s, when first Anna Bidder, again from Cambridge (there is quite a Cambridge tradition in this matter, and it is nice to feel part of it), and then Denton and Gilpin-Brown (from the Plymouth Marine Laboratory) went out to study feeding and digestion and finally settled the matter of how the buoyancy mechanism worked (see "Buoyancy," elsewhere in this book). Growth rates and longevity, the distribution of species, and many more details of the ecology of the animals have been established over the last twenty years by two geologists, Bruce Saunders and Peter Ward (who has written a very readable account of all this history); I was lucky enough to join them on one of these trips, to Lizard Island on the Great Barrier Reef, in 1985.

This story illustrates, I think, one of the nice things about my trade. As I have been fortunate enough to know it, it is almost a family business. In the less fashionable areas, at least, the protagonists all know each other. If they are in fact in competition, this is generally swamped by a mutual enthusiasm for what they are doing. Long may it last. I suspect it won't. There is too much money in cells and molecules, and money is a very poisonous commodity.

My wife and I joined the fellowship of nautilus hunters in 1984. At that time nobody had ever caught one along the south coast of New Guinea. But they had to be there; the ecology was perfect, and shells were sometimes found along the strand line at Motupore. So we bought quantities of cheap rope and made a trap from mangrove poles and chicken wire, baited it with dead fish, and set it in six hundred feet of water, three miles out to sea, off the barrier.

Getting it up next day was the problem. The fourteen-foot dory, a flat-bottomed aluminum tray with an outboard motor, had no winch, and it is physically almost impossible—for me anyway—to haul six hundred feet of rope with a heavy weight at the end into such a boat without, at least, some sort of ratchet. The solution was quite simple; you need two such boats, and happily we had two at Motupore. To one you attach a pulley; the trap line is passed through this and tied to the stern of the other. Take off in opposite directions. This is not an exercise to commend itself to a health-and-safety-at-work inspector—the boats swamp if pulled backwards—but it can be done, given a flattish sea, frantic hand signals between the two boats, and a bit of luck.

The first three animals we caught died within hours. I hadn't realised the desperate importance of keeping them cool. By the next weekend break I had scrounged a portable refrigeration unit for the lab and an insulated picnic box to take out in the boat, and the animals survived perfectly. They fed like starving hyenas, fornicated like rabbits, and spent most of their time sitting quietly and doing nothing, a condition we now suspect is typical of their life in the sea.

Opportunity knocked. We had an inaccessible animal in a remote place and my traveling kit of small apparatus that would work off car batteries. In the next few weekends we were able to record oxygen uptake (compared with the octopuses that we normally work on they were very slow-living animals) and examine ventilation—the means by which they pass water through the gills to collect the oxygen they need. Crude experiments—such as cutting windows in the shell and using skimmed milk to make the flow visible—took place alongside the more sophisticated use of pressure transducers and impedance-change measurements to tell us which bits of the body were moving in what sequence. It is easy to become an instant expert if you turn up in the right place at the right time.

And quite suddenly you begin to remember that the beast you are looking at is really old. Animals very like *Nautilus* have been around for three or four hundred million years. Long before mammals, long before reptiles, even long before fish as we know them, animals like this were swimming in the ancient oceans. What you have in your hand is history, something unbelievably ancient; this is the way one sort of life once was. Or so you can hope. The shells were like this; you can never be quite sure about the soft parts, because soft parts rarely leave convincing fossil evidence. But if the shell stayed the same, and the scars left by the muscle attachments to the shell look the same, it is not unreasonable to guess that the muscles were laid out in the same way and that the animal still moves and breathes in the manner of its ancestors. It is, in any event, the only model we have. A physiologist, in the absence of any evidence to the contrary, can legitimately believe that he is examining a fossil physiology as well as an ancient morphology.

Provided he is wary. All the other shelled cephalopods, the nautiloids and the ammonites that populated the seas in Palaeozoic and Mesozoic times, are extinct. *Nautilus* alone survives. It could have been a fluke. It is more likely that it survived because it could do something that the rest could not.

On our second trip to Motupore, in 1989, the people on the island pulled out all the stops to provide refrigeration. First they produced a freezer with a hose running through it. But this proved a dud because the aged, paraffin-powered, freezing mechanism had died. Then we tried using the guts of an electric domestic refrigerator to cool a jacket round a tank, with our seawater supply trickling through it into a big aquarium. That worked, wrapped in all the polystyrene packing we could glean from the university. And, finally, a six-foot cube, an insulated box with a door and shelves inside, once a shop's cold store, that now became our stockroom, with four aquaria for the nautiluses. It came over to the island precariously balanced on one of the dories, and was carried triumphantly into the lab by a dozen large highlanders amid much noise and confusion. They stayed for the rest of the day, drinking our beer and sleeping off the excitement.

By this time we knew, more or less, how to trap the animals. Trial and error had shown that buoys off the reef tended to vanish. The tidal streams are very strong and the rope is too valuable to expect poor fishermen to leave it alone. So this trip I dived and tied the trap line to the reef face in thirty or forty feet of water and motored out to sea with our third of a mile of cable before tossing the trap overboard to settle in six or seven hundred feet; the drop-off is very steep. Recovery meant some quite accurate plotting with a sighting compass and another dive to pick up the rope. There are some professional advantages in being a yachtsman, and any excuse to dive on the outer barrier was not to be ignored; even if the water was sometimes a bit murky for photography, the crayfish would be marvelous for supper.

We never failed to collect *Nautilus.* Sometimes we got ten or twenty. Long may that last. So far we were the only people fishing there. In the Philippines, there is a commercial fishery, providing shells for mantelpieces. In 1971, it was possible for one woman to sample three thousand of the animals from the Tanon Strait with three traps. In 1987, Peter Ward revisited the area. In a week, ten boats with forty traps caught three, perhaps the last three, animals. It is easy to kill off a scavenger that homes in on traps. The tragedy is that *Nautilus* is not the only animal that is being obliterated in this way; any large, slow-growing invertebrate with a decorative shell is vulnerable, and many of the large marine snails are becoming rare. They look beautiful in shell shops. Don't buy them.

Enough of preaching. What do we know about *Nautilus,* the living fossil that we are doing our best to extinguish? (All right, I kill them too. But not very many, and at least we learn something before the shells end up in a museum or my sitting room.)

Nautilus is Indo-Pacific. It could probably live perfectly well in the Caribbean, but continental drift ensured that the Atlantic opened up only after the isolation of *Nautilus* as the only survivor of the shelled cephalopods, and very cold water apparently stops it getting round South America or South Africa, let alone the Northeast Passage. Nor is it likely to benefit if the Panama Canal is blasted down to sea level and becomes a saltwater freeway. There seem to be no nautiluses east of Fiji. Isolated patches of suitable habitat are cut off by deep water because *Nautilus* does not have planktonic larvae and cannot penetrate below about five hundred meters without imploding. Surface temperatures are lethal to it, and it gets hammered by fish and turtles as soon as it ventures into the top fifty meters. So it is limited to long narrow strips, off reefs bordering on deep water.

Within the Indo-Pacific, wherever it *is* found and unexploited, *Nautilus* is a common and successful animal. There are five recognized species and one regarded as questionable; it may just be a very large variety of one of the others. Live specimens are found as far north as Japan and as far south as Western Australia. Mark, release, and recapture experiments, carried out by Bruce Saunders in Palau, show that individuals can wander several kilometers in a day, along the reef face. Marked shells float ashore hundreds of miles from the point of release, and this may give the impression that it is more widespread than is really the case. The status of *Nautilus* off East Africa, for example, is unknown. The nearest live location is the Andaman Islands. Occasional shells are found off the Kenyan coast, presumed to have drifted clear across the Indian Ocean. The Kenyan government put out a postage stamp showing one; they spelt it wrong, *N. pompileus,* with an *e* instead of a second *i*; now quite a collector's item. But who knows? Is the animal not there, or is it simply that nobody has seen fit to put down a thousand feet of rope to check? It would be fun to try to find out.

Nautilus is long-lived. Mark and recapture in the wild, studies of animals kept in aquaria, and analyses of bits of shell from wild-caught animals (the proportions of short-lived radioisotopes change with age) all show that growth continues for ten to twelve years, after which the animals become sexually mature and lay a dozen or so massive eggs per year, probably for the next dozen years. Very odd, that is. All the other cephalopods that we know much about grow like smoke for a couple of years, breed once only, throwing everything into a brief reproductive holocaust.

Whether longevity was typical of the extinct shelled cephalopods will never be known for certain. The odds are against, at least for the ammonites, which constitute most of the fossils that we find. If you grind down fossil ammonites to look at the innermost whorls of the shell, you can tell, as you can with *Nautilus,* how big it was when it hatched. The infant shell is smooth, denoting regular growth, while the young animal is still within the egg, feeding on the yolk. The shell surface becomes much more irregular when the poor little beast hatches and has to fend for itself. *Nautilus* hatches out as quite a large animal, all of an inch across; ammonite babies measured millimeters. So the odds are that they were produced in much larger numbers. That, and the occurrence of great drifts of ammonite shells, all much of the same size, in many deposits, suggests a life cycle much more like that of the modern squid and octopuses, big-bang reproduction followed by an exhausted death.

Other aspects of the physiology of *Nautilus* are more likely to be typical of the lives of extinct forms, such as the manner in which the animal develops the jet that drives it along; with no fins and no clambering arms like an octopus, it is the only way that *Nautilus* can get about. It develops a jet in two different ways. One is by the rippling movements of the flaps formed by the hind edges of the funnel, which curls about to direct the jet. This is how the animal ventilates the gills, drawing water in close to the center of the spiral of the shell on either side and wafting it out through the spout of the funnel. The flow will blow the neutrally buoyant animal backwards at a few centimeters per second if it doesn't hang onto the substrate. If *Nautilus* wants to move faster, it can develop a much more powerful jet by drawing the head back into the shell, squirting out the water in the gill cavity. Pressures rise by an order of magnitude, as do speeds; a powerfully jetting *Nautilus* can top thirty centimeters per second.

No big deal. Fish of comparable size can do a lot better, ten rather than three lengths per second. But absolute speed may not be everything; what matters in the long run is the cost of locomotion, the fuel expended in getting from one patch of food to the next. That was one thing that we studied in 1989. A colleague from Canada, Ron O'Dor, who is another cephalopod nut and an old friend with whom I have worked, on and off, here and there around the world for the past twenty years, joined us at Motupore with an ancient Apple computer and a transducer/transmitter that could record and transmit the pressures generated within the shell when the animal was jetting, or simply breathing. Together we built a flume, a tank with a flow of water generated by one of us cranking a propeller while another shot videos of the animal swimming against the current, stimulated by the smell of dead fish in the water. Jet pressures were relayed by Ron's transmitter glued to the outside of the shell with a pipe into the funnel aperture. A remote hydrophone collected signals, relayed to the computer and synchronized with the video record—the sort of system that would have been quite impossible fifteen years ago, but is now transportable as cabin baggage.

Parallel experiments showed a strict relation between ventilation/jet pressures and oxygen uptake. That was important. Oxygen is the universal fuel. Some animals manage without it, gut parasites often have to, as do worms and other creatures living in foul mud. But the energy yield of anything they eat is rarely more than a seventh of what they get from oxidation, so animals are always aerobic if they can be. If they operate anaerobically for a while, they run up an oxygen debt, which must be paid off afterwards to eliminate all sorts of undesirable anaerobic by-products (the most familiar evidence of this is the way we get short of breath when we run, and have to stop and breathe deeply for a while afterwards). Given that we could translate pressure pulses into oxygen uptake, we had a means of costing anything that the animal happened to be doing, moving or sleeping or fornicating or digesting.

Ron's rather ancient computer—he is sentimental about it, like somebody running a vintage motorcar, but by '89 it really was time to get an update—could only store four minutes' worth of data at a time. So we settled for finding out the cost of locomotion, and, because we could plug the pressure transducer into the heart, for looking at the circulatory changes that accompany activity. What we found was predictable: the oxygen cost of swimming rises steeply with speed. We also found that the animal cannot sustain speeds above fifteen centimeters per second or so without running into oxygen debt. Locomotor costs at such speeds were two or three times higher than those of fish of similar size, a seemingly overwhelming disadvantage, until we also discovered that at very slow speeds, *Nautilus* actually travels more economically than fish. The secret is that while fish have to swim and breathe, *Nautilus* can just breathe, and gets wafted along on its ventilatory stream. This, maybe, was one secret of the success of the early shelled cephalopods, a reason that they managed to survive for so long in the teeth of sustained opposition from fish that were faster and much more economical over most of the speed range.

To take matters further we needed, of course, to see what *Nautilus* actually does in the sea, not what it can do when kept in a laboratory. Is it continuously, or only intermittently, active in the sea? Does it, in fact, use its ability to survive in low-oxygen conditions to explore pockets of low oxygen—if such exist—where there might be food inaccessible to its fishy or crustacean competitors? The transmitters are expensive, £250 or so a time, so in 1989 we never actually risked letting an animal loose. Instead, we lowered one in a cage, recording its heartbeat as it descended to the better part of one thousand feet. The transmitter worked a treat. Everything was now possible.

That was in 1989. In 1991, we returned to New Guinea; the experiments last time had worked so well that we had no particular trouble raising the money for a further expedition. This time we planned to release animals off the reef and follow them as they went about their business, tracking them from a boat hovering overhead with a directional hydrophone.

We had one amazing stroke of luck. There was a war in the Persian Gulf. Saddam Hussein invaded Kuwait. We must have been the only people outside the Kuwaiti royal family to derive any benefit from the Gulf War, and it happened because of the way mariners now plot the position of small boats. We had a GPS (global position from satellites) unit. These things work from U.S. military satellites and can fix a position, anywhere on earth, within centimeters. Except that the U.S.A. isn't that stupid. Letting people access this system would be tantamount to giving all terrorists a system for delivering bombs literally to any doorstep they cared to select. So the signal from the satellites is scrambled, just a bit, so that positions can only be plotted to within a hundred meters or so. Useful for yachtsmen, not quite good enough for homemade cruise missiles.

During the Gulf War, every tank commander had one. Except that because the allied armies were assembled in rather a hurry, there weren't enough descramblers to go round. So they shut off the scramblers for a bit. Bully for the *Nautilus* trackers; we were able to record a unique data set, plotting the movements of our animals (or, rather, the boat above the animals) within meters.

What the data showed was that *Nautilus* in its natural habitat is a very economical animal. It potters about very slowly, along and up and down the reef face, a cropful of food at this rate would last it for weeks. Most interestingly—and we might have guessed this from the shape of the shell and the nature of its buoyancy mechanism—it costs almost exactly the same to move horizontally as vertically, something that fish, with their need to inflate and deflate the swim bladder as they change depth, cannot do. This, then, was a possible reason why *Nautilus* was able to survive in the same habitat as fish, which, on the face of it, could swim circles round it and get to any available food in a fraction of the time. They could, by all means, but *Nautilus* could get there more cheaply.

That sounded too easy. There had to be more to it than that, and a possible answer came up when we took seawater samples from the places where *Nautilus* was living. They were notably oxygen deficient. We already knew that *Nautilus* was good at coping with low oxygen. There is no other cephalopod that anyone has studied that can be locked in a box overnight, with its oxygen reduced to vanishing point, and still revive as soon as oxygenated water is once again available. How was that done?

We went back again in 1993, this time with Bob Boutilier, a colleague from my department in Cambridge, Canadian, poached from Ron's department at Dalhousie. Bob is an expert on blood-oxygen transport and a biochemist, which I am not. He found that *Nautilus* has a blood-oxygen dissociation curve and ways of trimming its metabolism ideal for living in really low oxygen conditions. In the course of evolution, the animal has settled for a blood pigment that can take up oxygen from environments that would normally be regarded as undesirable, if not uninhabitable, by any self-respecting predator. The rate of uptake is inevitably slow, because there is very little oxygen to be scavenged. What *Nautilus* has hit upon is the idea of being intermittently active, tanking up on oxygen for an hour or more to sustain a few minutes of aerobic activity once in a while. It can remain an active predator in places where no active predator ought to be.

Maybe the extinct ammonites had the same ability. Any animal that must retreat into a shell when danger threatens has to be able to cope with intermittent low-oxygen conditions; their metabolism must in this respect have been similar to that of *Nautilus.* Maybe they, too, could store oxygen and dive into regions of low oxygen content. Maybe they, too, had evolved the trick of intermittent activity. Throughout the Palaeozoic and much of the Mesozoic there were plenty of opportunities to exploit such abilities in seas that were typically much less well oxygenated than now.

In which case, why did the ammonites die and *Nautilus* survive?

The reason may have had nothing to do with exploitation of a diminishing habitat, as all but the deepest, impenetrable oceans become progressively oxygenated. In the end, it was probably their life history that killed them. Huge numbers of planktonic larvae following a reproductive holocaust was the wrong strategy. Fish, and perhaps also crustacean and cephalopod predators, eventually became too quick for tiny shelled ammonites. A turkey shoot. Or maybe the passenger pigeon is a better model. At all events, they snuffed it.

Well, that's as good a theory as any. In science you are never proved correct. You only have the best explanation until a better one turns up. That's part of the fun of the game.

THE DILEMMA OF THE JET SET

My favorite animals, you will by now have surmized, are the cephalopods. Squid, cuttlefish, octopuses, and *Nautilus,* a living fossil that gives me an excuse to swan off to the Pacific and there spend a quite unnecessary proportion of my time underwater.

Cephalopods proceed by jet propulsion. It all began way back in the Palaeozoic, when fish were still a gleam in the Creator's eye and some small, limpetlike creatures chanced upon the curious capacity to accumulate gas in the apex of their shells. No big deal, you may surmise, flatulence can happen to anybody. But in the early days of the Palaeozoic, the ability to generate and hang on to gas (it wasn't flatulence, for reasons explained in another article; see "Buoyancy," elsewhere in this book) was important. It made the little limpets light on their feet. When danger threatened, the usual molluscan defense (a medieval tactic—pull back into the castle and slam the door) resulted in a very unusual response. The squirt of water expelled as the body pulled back into the shell would have pushed the little animal off the bottom, a leap into the blue that could have been totally baffling to would-be predators, thrashing and gnashing about on the seafloor—few if any animals of any size had solved the problem of midwater swimming before the end of the Cambrian. Control could be improved by rolling up a part of the foot to form a funnel, directing the squirt of ejected water.

A great invention, propulsion by jet, the ultimate defense against flatlanders, a world safe for protocephalopods.

And unsafe for everybody else. Because the leaping defense also formed the basis for a leaping attack. If neutral buoyancy could make possible a speedy withdrawal in the face of serious opposition, it could also provide the means for dropping like a hawk from above on the other members of the fauna bumbling along on the bottom.

The early cephalopods rapidly became the dominant predators of the mid-Palaeozoic seas. They multiplied exceedingly, diversified, and produced forms ranging from a few centimeters to several meters long. No doubt they ate one another. They certainly made life difficult for everybody else, dropping unannounced from above, withdrawing into inaccessibility with some poor victim clasped in the tentacles formed from the once-creeping foot. Hawks among the mice, for millions of years the cephalopods had it all their own way.

Nautilus is interesting because it has a shell that closely resembles the fossil shells of some of these early forms. It is a reasonable bet that the soft parts have also stayed very much the same; we can trace some features of the musculature from the scars in the fossil shells, and we know a bit about the hard mouthparts. We cannot, of course, be sure about things like the heart or the gills, but even so it is well worth looking to *Nautilus* as an indication of how things were in that very distant past. It is the only clue we have.

Problems hit the early nautiloids later in the Palaeozoic. Fish appeared, clumsy creatures at first, but rapidly evolving swift, undulating bodies, and later their own forms of neutral buoyancy. Fish were maneuverable, and fish had jaws.

Fish also had a locomotor system that was just as fast and inherently much more economical than that of the jet set. Consider the elementary physics of the matter. Thrust, to drive an aquatic animal along, depends on the volume and the velocity of the water pushed backwards, by the sweep of a tail or in a jet ejected from a body cavity. Thrust equals mass times velocity. So the same thrust, pushing the animal along, can be achieved by pushing back a liter of water at ten centimeters a second, or two liters at five, or half a liter at twenty. The crunch is accelerating the water to the required velocity. $E = \frac{1}{2} mV^2$, another fact of life. In energetic terms it costs four times as much to drive the half liter back at twenty centimeters as the two liters at five.

Bad news for the shelled cephalopods. The ejectable mass, squirted out by pulling back the head into the gill cavity, was limited by the shell. Fish had no such constraints. What mattered to them was the area swept by the tail, which could easily be greater than the cross section of the body, pushing back a relatively enormous volume of water. With a relatively leisurely wiggle, a fish could achieve the same thrust as a similar-sized cephalopod squeezing desperately to accelerate a much smaller volume to a much greater velocity. The fish system was more economical, and potentially much faster. The propulsive system that had emerged as a winner in the Palaeozoic proved to be something of a white elephant in the Mesozoic.

There was not a lot the shelled cephalopods, ammonites and all the rest, could do about it. They struggled for a long time, streamlining the shell, reducing the weight of the shell (90 percent of the buoyancy in *Nautilus* is squandered on supporting the shell), and developing some very fancy architecture to maintain rigidity in a sometimes paper-thin structure. But in the end the fish finished them. *Nautilus,* a deep-water form adapted to survive on the fringes of barrier reefs in conditions that may be unsuitable for most predatory fish, is the only one we have left.

Other cephalopods survived by scrapping or enclosing the shell. The only way they could increase the ejectable mass was to free the gill cavity from the constraint of a rigid box. Cephalopods grew forwards, extending beyond the shell. The wall of the gill cavity, hitherto the lining of the shell, became muscular, capable of expansion and contraction, sucking in and pushing out a much larger volume of water than had ever been possible with the nautiloids or the ammonites.

A few cephalopods retained the shell. The white cuttlebones that you find on beaches in the Pacific and on the eastern side of the Atlantic are the shells of the cuttlefish, *Sepia,* a cephalopod with a body that encloses and hangs below the source of buoyancy. If you look at it carefully, you can see that the cuttlebone consists of a huge number of laminar air-filled chambers—it is said to form one for every day of its life—a clever adaptation for getting the maximum buoyancy for the minimum weight. Beachcombers in the tropics will also be familiar with the tiny coiled and chambered shells of *Spirula,* a deep-water, open-ocean animal that carries the shell enclosed in the tip of its abdomen.

But these are the exceptions. Most living cephalopods, the squid and the octopuses, have reduced the shell to a mere vestige, serving to stiffen parts of the otherwise floppy body. It no longer contributes anything to the animal's buoyancy.

Abandoning neutral buoyancy was undesirable because without it an animal must swim continuously upward to remain in midwater. But it permitted much more extreme streamlining, reduced drag coefficients, and generally allowed the squid to achieve speeds and maneuverabilities comparable to those of their fishy competitors.

Scrapping the external shell also meant, of course, abandoning the traditional medieval defense. Squid had chucked away their armor; if danger threatened, they now had to cut and run; it was a considerable help that they could now run a lot faster. As the cephalopods progressively reduced the shell, the arms race with fish moved into a new phase. It began to depend on observation and anticipation, on better eyes and better brains, rather than on an old-fashioned capacity to stand and take punishment. Dreadnoughts became frigates. Cephalopods have marvelous eyes and bigger brains, size for size, than fish. Midwater squid took to shoaling, like fish, for mutual protection; a hundred pairs of eyes are better than one. Others, the octopuses, took to skulking around the rocks like groupers or blennies, while yet others, the cuttlefish and sepiolids, took to a life close to the bottom, like flatfish, and frequently buried out of sight within it, the eyes alone protruding from the sand, alert and unblinking.

A host of abyssal forms came to parallel some equally hideous fish. Some of these, living in dark, food-scarce environments, returned to neutral buoyancy. Having abandoned the shell, so that a gas chamber was no longer an option, they reduced their weight in seawater by alternative means, trading heavy salts in their makeup for lighter materials, even, in some cases, for fats (see "Buoyancy," elsewhere in this book). Neutral buoyancy cut locomotor costs, but restored all the problems of bulk and slower speeds, so that two distinct strategies must have evolved. As well as the typically fast active cephalopods that we know most about, there is a host of relatively unknown slow-moving flabby creatures, largely inaccessible in the deep oceans. We don't know how long these forms live or how fast they grow.

What we do know is that all the shallow-water forms that we catch and eat are flash-in-the-pan animals, growing fast and living for year or two only before they breed once and die. Arguably, this pattern has been forced upon them in the past by the high running costs of jet propulsion. They could compete with fish so long as they worked flat-out, dodging the opposition and preying upon food that was seasonally plentiful. What they were not so good at was coasting along in times of relative scarcity. A large cephalopod was therefore likely to contribute more to the survival of the species if it spent itself spawning at the end of a rich summer rather than if it hung about, melting away during the winter; the typical life history of the squid and cuttlefish that we know about lasts less than two years.

There was, of course, one further trick to pull. Along with the adoption of fishy lifestyles, many cephalopods evolved fishlike means of propulsion. The lateral fins, developed, one must suppose, in the first instance, as control surfaces, began to assume a role in swimming, beating like the wings of a skate. Figures for the cost of this sort of transport are only slowly becoming available, but it seems inescapable that a mainly fin-propelled squid will prove more economical if somewhat slower than a squid propelled by an extravagant locomotor jet.

The jet will persist, whatever the preferred means of swimming. An animal with a directable jet can do things that a fish cannot. In an emergency, it makes it almost as fast forwards as backwards, a difficult prey for a would-be predator. The jet can be used to dig holes to settle in, to flush out shrimps and other food, and even to aerate the eggs (octopuses brood theirs). And it can be used as a gun to discharge the cephalopod's secret weapon, ink. (And it literally is ink, used in the past for drawing and writing; it is no accident that *Sepia* is the Latin name of the cuttlefish, the source of ink.) The ink is secreted into a sac and can be discharged at will to produce a smoke screen or, more subtly, it can be entangled in a ball of mucus, as a decoy to distract the enemy while the cephalopod escapes.

But, as the engine, the jet is competitive only at the extreme ends of the speed scale: flat out, when cost is a secondary consideration, and, perhaps rather surprisingly, in very slow locomotion, which is only possible in forms with neutral buoyancy. At this end of the scale the ventilation stream, necessary anyway to bring oxygen to the gills, propels the animal, and here, for once, fish seem to lose out. A slow-moving fish must pump water through the gills by means distinct from its normal propulsive machinery, and that, it seems, runs it into minimal costs in excess of the breathing/propulsive jet of the cephalopod. This, so far shown only for *Nautilus* (itself, one might note, the only slow-growing, long-lived cephalopod that we have so far recognized—they live for ten or twenty years instead of the usual two), is perhaps the secret of the success of many of the abyssal forms which have achieved neutral buoyancy by retaining ammonia (again, see "Buoyancy"). For the rest, the midspeed swimmers, there is no option but to go for fishlike, fin-driven locomotion. If you can't beat them, join them. For most cephalopods, it is the only solution to the dilemma of the jet set.

DOES SCIENCE HAVE TO BE USEFUL?

Biologists are scientists, and traditionally regarded by physical scientists and mathematicians as belonging to the soft end of the spectrum, hard, or more strictly mathematical, science being the real thing. Biologists, say real scientists, are generally pretty competent observers, but inclined to anecdote and weak on quantification. They exhibit a deplorable tendency to hedge when it comes to conclusions, are liable to insist that most cats are grey, when a rigorous hard scientist would soon have sorted out that they were either black or white. People who have to make decisions between alternatives, politicians and captains of industry, good arts men for the most part, knowing in their hearts that most questions are not answerable on a yes-no basis, nevertheless consult real scientists when they need answers and prefer to believe advisors brought up in the tradition that there has to be a precisely correct answer to every properly framed question. That, says conventional wisdom, is what science is all about.

The fact that everyday experience shows that this is nonsense is in a large measure responsible for the dichotomy between the men of letters and the men of numbers. It provides an excuse for nonscientists to wash their hands of the whole business, to accept that science is inaccessible, and to believe that the way in which scientists regard the world outside their laboratories, if indeed they do regard it, is by and large irrelevant to the vast majority of mankind. Useful people if you want a word processor, an airplane, or an atom bomb, but irrelevant to the way most people interact and think about the world in which they live.

I am a university teacher, a zoologist. I think part of my job is to convince students that we do not simply study animals to acquire a technology. It will by all means help us to control the world around us, to exploit and conserve, and to find out, albeit indirectly, more about the behavior and physiology of our own species. More important is that the study of biology teaches a particular sort of approach to these problems, an acceptance that in most cases you are unlikely to come up with an absolute answer, unless you are prepared to frame questions in a manner that is unrelated to reality. Biological phenomena tend to be very complex, and one will rarely, if ever, be in possession of all the necessary information to completely explain anything. What one has to accept is an answer that is the best available in the circumstances. With luck, you can quantify the uncertainty. Biology teaches you to understand statistical statements. A biologist will say that he is 95 percent (or 99 percent) certain, whereas a politician or other salesman would be obliged to express total conviction. We are by training more honest than most.

This, in turn, profoundly affects the attitude that biologists have towards people. We know that we are fallible, and that all that competence does is to improve the chances of somebody's being right in his judgments. A hard scientist is more liable to believe that the inability to come up with a definitive answer is simply a sign of incompetence. I would hold that ours is the better training. In a real world full of maverick people and animals, some pretty bloody-minded plants, machinery regulated by sod's law, and a God almighty whose track record for arbitrary intervention is well documented even in his own publications, to pretend to be absolutely certain is to invite disaster. If you want a realistic assessment, ask a biologist. You may not like the answer, but at least you won't be kidding yourself. Captains of industry, politicians and planners, please note.

A second reason for the study of biology is quite simply that it will make the world a far more interesting place in which to live. Besides being a marine biologist—my particular field of research is the physiology of cephalopods, a group that includes squid and octopuses—I am a gardener and a painter, a SCUBA diver and a yachtsman. All these are areas in which a knowledge of plants and animals makes the fine grain of my surroundings more entertaining—and, just occasionally, safer. I have never met a bored biologist, at least not among the sort who study whole animals or plants (an ever-increasing proportion of us are chemists or physicists, and as a whole-animal man I cannot pretend to speak for them, though all my contacts to date suggest that there is nothing fundamentally wrong with them). Biologists suffer from paranoia, frustrated ambition, angst about their sex lives, lack of hard cash, and all the usual frets that beset mankind. But they are not bored.

This last is no trivial reason for the study of animals. Alleviating what might otherwise prove to be a monotonous existence is one reason we read books, watch plays, listen to music, or go to exhibitions. Nobody claims that the arts have to be useful. It is sufficient that they make life more interesting. The notion that science should necessarily be useful is one of the great con jobs of the second half of the twentieth century, perpetrated by governments that realise that science can sometimes prove outrageously expensive. This is no good reason for deciding that research always has to carry the potential of paying for itself. It can also be very entertaining, and not only for the protagonists. One has only to note the popularity of natural-history and other science programs on the television, the media cashing in and skimming the cream from an enormous background of hard-won and often quite useless knowledge. The fact that apparently useless studies, blue-sky research, not infrequently proves to be useful later is a bonus; it should never be the sole reason for embarking upon any investigation. Animals, like any other exhibition—and this one is generally for free—become interesting in proportion to what you already know about the subject, and a part of the job of any zoologist is to help this process along. The fact that he may be getting more fun out of making his contribution than any of his audience is not a reason for supposing his research to be selfish, an ivory-tower activity unrelated to the work, wealth, and happiness of the rest of mankind. A writer, a painter, or a musician is in a very similar position. Some of their products are quite as inaccessible, explorations at least initially of interest only to a few fellow members in the trade. We tolerate, even subsidise, their activities because we believe that they increase the range of experience available to other people. Science is like that, too.

All this sounds very defensive. It is. Sometimes I feel just a little guilty. Many years ago the man who held the job that I now have warned me. "You have to face it," he said, "you will never be a rich man if you stay in research. But" (and he paused), "it is a very remarkable world that is prepared to pay a man to play for the whole of his life."

INDEX